Optimization of Micro Processes in Fine Particle Agglomeration by Pelleting Flocculation

T0179123

Examination Board

Chairperson
Prof. Dr. rer. nat. habil. Marion Martienssen
Department of Biotechnology for Water Treatment
Brandenburg University of Technology Cottbus-Senftenberg

Supervisor
Prof. Dr.-Ing. habil. Peter Ay
Department of Minerals Processing
Brandenburg University of Technology Cottbus-Senftenberg

Other members
Prof. Dr. rer. nat. habil. S. Narra
Department of Technical Environment
University of Applied Sciences, Lübeck

Prof. Dr. rer. nat. habil. Uwe Harlander
Department of Aerodynamics and Fluid Mechanics
Brandenburg University of Technology Cottbus-Senftenberg

Optimization of Micro Processes in Fine Particle Agglomeration by Pelleting Flocculation

A thesis approved by the Faculty of Environmental Sciences and
Process Engineering at the Brandenburg University of Technology in
Cottbus in partial fulfilment of the requirement for the award of
the academic degree of Doctor of Philosophy (Ph.D.) in Environmental Sciences

by

Master of Science
Benjamin A. Oyegbile

from Ibadan, Nigeria

Supervisor: Prof. Dr.-Ing. habil. P. Ay
Supervisor: Prof. Dr. rer. nat. habil. S. Narra
Date of oral examination: 04.02.2016

Cottbus

February 2016

CRC Press
Taylor & Francis Group
Boca Raton London New York Leiden

CRC Press is an imprint of the
Taylor & Francis Group, an **informa** business

A BALKEMA BOOK

Optimierung von Mikroprozessen bei der Agglomeration Feiner Partikel mittels Flockungspelletierung

Dissertationsschrift zugelassen von der Fakultät für Umweltwissenchaften und Verfahrenstechnik der Brandenburgischen Technischen Universität Cottbus zur Erlangung des akademischen Grades „Doctor of Philosphy (Ph.D.) in Environmental Sciences"

von

Master of Science
Benjamin A. Oyegbile

aus Ibadan, Nigeria

Gutachter: Prof. Dr.-Ing. habil. P. Ay
Gutachter: Prof. Dr. rer. nat. habil. S. Narra
Tag der mündlichen Prüfung: 04.02.2016

Cottbus

Februar 2016

CRC Press
Taylor & Francis Group
Boca Raton London New York Leiden

CRC Press is an imprint of the
Taylor & Francis Group, an **informa** business

A BALKEMA BOOK

Published by: CRC Press/Balkema
P.O. Box 11320, 2301 EH Leiden, The Netherlands
e-mail: Pub.NL@taylorandfrancis.com
www.crcpress.com – www.taylorandfrancis.com

First issued in paperback 2020

© 2016 Taylor & Francis Group, London, UK
CRC Press/Balkema is an imprint of the Taylor & Francis Group, an informa business

No claim to original U.S. Government works

ISBN 13: 978-0-367-57472-7 (pbk)
ISBN 13: 978-1-138-02861-6 (hbk)

Typeset by Quick Sort (India) Pvt Ltd., Chennai, India

Library of Congress Cataloging in Publication Data

"It is some fundamental certainty which a noble soul has about itself, something which is not to be sought, it is not to be found, and perhaps, also, is not to be lost – The noble soul has reverence for itself."

Friedrich Nietzsche

Dedication

To the One Who Left Too Soon

Declaration

I Benjamin A. Oyegbile hereby declare that this PhD thesis was completed and written independently of any unauthorized help at the Brandenburg University of Technology, Cottbus, Germany within the framework of the International PhD Programme ERM. All used sources were duly cited in the thesis and included into the list of references. The PhD thesis has been completed under the supervision of Prof. Dr. Peter Ay and Prof. Dr. rer. nat. Satyanarayana Narra.

This PhD thesis has never been submitted elsewhere in part or in whole for a degree at any other institution.

Benjamin A. Oyegbile
(Matr. No.: 3242702)

Abstract

This study addresses the problem of low throughput in the technical application of shear-induced agglomeration in Taylor-Couette devices, where particle aggregation is achieved by pelleting flocculation. Pelleting flocculation, an offshoot of classical orthokinetic flocculation, is an effective structure formation process based on the "metastable state" concept, and has been recognized as one of the most promising wet agglomeration techniques. The mechanism of agglomeration by pelleting flocculation is by the exudation of the dispersion medium (pore water) through the application of fluctuating uneven forces on the floc surface. This is realized through the rolling and collisional effects induced by the propagated vortex flow as well as the geometry of the agglomeration vessel.

In this investigation, a new pre-treatment technique was developed as a technical proof of concept based on Taylor-like forced vortex flow structure propagated within a rotor-stator cavity in order to induce rotational and collisional effects. A simulation of this processing technique was performed experimentally by shearing a model suspension in bench-scale vortex reactor units, with a view to optimize the micro-processes that are vital for the structure formation. Polydispersed synthetic substrates were prepared from kaolin and ferric hydroxide powder with median particle sizes of 5.07 μm and 6.14 μm respectively dispersed in distilled water. Polyacrylamide—PAM (Superfloc®N-300) and quaternized cationic acrylamides—DMAEA-Q (Superfloc®C-491 and Superfloc®C-492) were used in sequential combinations as bridging materials.

Using a non-conventional stirrer-vessel system (reactor geometry and stirrer configuration), the micro-processes were optimized for a number of operating conditions by the means of quasi factorial experimental design. The hydrodynamic characteristics of the turbulent flow in the reactor and the vortex pattern of the flow stream were characterized by computing the time-averaged velocity profile and vorticity map using Particle Image Velocimetry (PIV).

The experimental results show that increasing the shear rate improves the efficiency of the pre-treatment process in the vortex reactors. However, a critical shear rate of 6811 s^{-1} and 7681 s^{-1} and kinetic energy dissipation rate of 42.49 and 54.05 m^2s^{-3} were observed in the batch and continuous reactors respectively after which there was a significant deformation of the pellet flocs. The solids content of the pellets flocs and the particle removal efficiency R_x were monitored in the pelleting process. The data from this study show that the dewatered pellets obtained from the batch reactor are nearly spherical in shape and exhibit a narrow size distribution with a maximum mean solids content of approximately 30% and a removal efficiency of 97.5%. The average pellet sizes are 3.7582 and 3.8460 mm with a mean compressive strength of 0.4298 and 0.4351 Nmm^{-2} for rotation speed of 145 and 165 rpm respectively.

It is anticipated that in the future, a numerical simulation of the flow and the pelleting process in these reactors can be implemented in order to achieve a higher level of process optimization and control.

Keywords:
Pelleting flocculation, turbulence, hydrodynamics, micro processes, shear reactors

Table of contents

Acknowledgements

The author would like to use this opportunity to thank everybody that contributed to the successful completion of this dissertation work.

First of all, I will like to thank God for making this undertaking possible. I will also like to express my profound gratitude to my advisor Prof. Dr.-Ing. habil. Peter Ay; Chair of Minerals Processing for his support, mentorship and valuable critique which contributed immensely and helped in shaping the focus of this research work. I am thankful to him for bringing his vast experience to bear on this study.

I am also grateful to Prof. Dr. Dr. h.c. Michael Schmidt; Dean, Faculty of Environmental Sciences and Process Engineering, Prof. Dr. rer. pol. Frank Wätzold; Chair of Environmental Economics, Prof. Dr. rer. nat. habil. Manfred Wanner; Chair of Ecology, and Prof. Dr. rer. nat. habil. Satyanarayana Narra of Lübeck University of Applied Sciences for their mentorship, help, advice and critique throughout the period of this research work.

I also acknowledge the help and support of Ms. Claudia Glaser and Ms. Grit Gericke of the Chair of Minerals Processing for their practical and administrative assistance. A special thanks goes to our technical support staff, Mr. Uwe Kränsel at Campus North for his practical advice. A warm appreciation to Prof. Dr. rer. nat. habil. Uwe Harlander and Mr. Michael Hoff of the Chair of Aerodynamics and Fluid Mechanics for their contributions towards the PIV measurements and data analysis.

Lastly, my sincere appreciation to my family and colleagues for their support and encouragement over the years. Special thanks to my friends George Forji Amin and Melvin Guy Adonadaga for their intellectual inspiration and not forgetting my fiancée for her understanding, support and words of encouragement.

List of tables

List of figures

List of symbols

Symbol	Quantity	Units
$\bar{\varepsilon}, \varepsilon$	Mean or local dissipation rate of kinetic energy	(Nms^{-1}kg^{-1} or m^{2}s^{-3})
N	Particle number concentration per unit volume	(m^{-3})
N_o	Initial particle number concentration per unit volume	(m^{-3})
N_A	Particle number concentration with sizes below $d_{F,max}$	(m^{-3})
N_B	Particle number concentration with sizes above $d_{F,max}$	(m^{-3})
N_F	Floc number concentration per unit volume	(m^{-3})
C	Coefficient of floc strength	(-)
f_o	Orthokinetic collision efficiency	(-)
f_p	Perikinetic collision efficiency	(-)
r	Particle radius	(m)
d_F, d_p, d	Floc or particle diameter	(m)
$D_{F,max}, d_{F,max}$	Maximum floc diameter	(m)
d_1, d_2	Longer and shorter pellet diameter	(mm)
d_{av}	Average pellet diameter	(mm)
d_{eq}	Diameter of an equivalent sphere of the same volume	(mm)
d_{10}	10 percentile equivalent diameter floc size	(µm)
d_{50}	50 percentile equivalent diameter floc size	(µm)
d_{90}	90 percentile equivalent diameter floc size	(µm)
d_A	Agitator or stirrer diameter	(m)
r_w	Eddy size or radius	(m)
β	Rate constant	(-)
α	Ratio of characteristic inverse times	(-)
α_{ij}	Collision efficiency factor	(-)
k, m, n	Exponents	(-)
η, μ	Dynamic viscosity	(Nsm^{-2})
K	Bond destruction coefficient	(-)
V	Stirring vessel volume or fluid volume	(m^3)
I_{SV}	Sludge volume index or specific volume of sediment	(mL/g)
P, P_{ave}	Power input or average power consumption	(W)
T_q	Shaft or mixer torque	(Nm or kgm^{-2}s^{-2})
N, n	Speed of agitator	(s^{-1}, m^{-1})

N_p	Dimensionless power number	(-)
V_c	Critical fluid breaking velocity	(ms^{-1})
V_{tip}	Agitator tip velocity	(ms^{-1})
V_θ	Flow tangential or circumferential velocity	(ms^{-1})
V_r	Flow radial velocity	(ms^{-1})
V_{pkol}	Peripheral velocity of vortex at Kolmogorov length scale	(ms^{-1})
v	Kinematic viscosity	$(m^2\ s^{-1})$
V_s	Sludge volume	(m^3)
ρ_T	Slurry solids concentration	$(g\ L^{-1})$
T	Absolute temperature	(K)
$\overline{G}$	Absolute or R.M.S velocity gradient	(s^{-1})
G_{ave}	Average or mean velocity gradient	(s^{-1})
G_L	Local velocity gradient	(s^{-1})
σ_{GL}	Variance of local velocity gradient	(s^{-1})
u	R.M.S velocity component	(ms^{-1})
ξ	Vorticity	(s^{-1})
ζ	Zeta Potential	(mV)
U	Pair bonding energy	(J)
p	Flocs or agglomerate porosity	(-)
Λ	Integral length scale of turbulence	(m)
λ_0	Kolmogorov micro scale of turbulence	(m)
λ_B	Batchelor micro scale of turbulence	(m)
λ_T	Taylor micro scale of turbulence	(m)
R_e	Dimensionless Reynolds number	(m)
ω	Angular velocity, speed or frequency	(rad/s)
ρ, ρ_f	Fluid or floc density	$(kg\ m^{-3})$
t	Time	(s)
R_x	Removal efficiency	(%)
V_T	Total flocculation energy barrier	(J)
V_A	Energy barrier due to Van der Waals attraction	(J)
V_R	Energy barrier due to Electrostatic repulsion	(J)
F_H	Hydrodynamic breaking force	(N)
F_{Hmax}	Maximum hydrodynamic breaking force	(N)
F_D	Hydrodynamic drag force	(N)
F_B	Aggregate cohesive or binding force	(N)
F_{max}	Maximum compressive force	(N)
γ_F	Pseudo-surface tension force	(N)
τ	Aggregate binding or cohesive strength	(Nm^{-2})
$\in$	Nominal axial strain rate	(s^{-1})
τ_c	Aggregate compressive strength	(Nm^{-2})
τ_s	Aggregate shear strength	(Nm^{-2})
τ_t	Aggregate tensile strength	(Nm^{-2})
τ_y	Aggregate surface shear yield strength	(Nm^{-2})
σ	Global hydrodynamic stress	(Nm^{-2})
σ_s	Turbulent hydrodynamic shear stress	(Nm^{-2})
σ_t	Turbulent hydrodynamic tensile stress	(Nm^{-2})
σ_y	Turbulent hydrodynamic tensile yield stress	(Nm^{-2})

List of abbreviations

DLVO	Derjaguin-Landau-Verwey-Overbeek
R.M.S.	Root mean square
CMC	Sodium carboxymethylcellulose
2-D	Two dimensional
3-D	Three dimensional
CFD	Computational Fluid Dynamics
PBM	Population Balance Model
NTU	Nephelometric Turbidity Unit
Poly-DADMAC	Polydimethyldiallylammonium chloride
Pes-Na	Polyethenesodium sulphonate
PCD	Particle Charge Detector
PAM	Polyacrylamide
DMAEA-Q	Dimethylaminoethyl acrylate—Quaternized
DMAEA-MeCl	Dimethylaminoethyl acrylate—Methyl chloride
MAPTAC	Methacrylamidopropyl trimethylammonium chloride
PIV	Particle Image Velocimetry
PTV	Particle Tracking Velocimetry
LDV	Laser Doppler Velocimetry
LDA	Laser Doppler Anemometry
LIF	Laser-Induced Fluorescence

Chapter I

Background, problem statement and outline

1.1 INTRODUCTION & STUDY BACKGROUND

The effective management of sludge and residuals which are by-products of many process industries such as minerals processing, water & wastewater treatment, pulp and paper, petrochemical etc. is very important from an environmental and economic perspective. There is a growing awareness about the technical and economic problems associated with residuals' processing and the realization that improvement in sludge processing can serve as a tool for improving environmental quality, as well as natural resource conservation, through water re-use, nutrient recycling and biomass utilization. Increasing sludge production and the energy-intensive nature of the current treatment processes makes it imperative to seek more innovative and efficient sludge dewatering techniques.

Management of the solid and concentrated contaminants removed by various industrial treatment processes remains perhaps the most difficult and elusive of the contemporary environmental engineering challenges (Biggs, 2006; Mahmoud et al., 2013; Tchobanoglous, Burton, & Stensel, 2003; Vesilind, 1988). Sludge is often characterized by high water content (low suspended solids concentrations) and high resistance to dewatering. Solid-liquid separation prior to disposal or utilization is therefore required to reduce the cost of transportation or comply with regulatory requirements, and this currently accounts for substantial part of the operational cost of many process industries (Mowla, Tran, & Allen, 2013; Yuxin et al., 2014). Though incremental progress has been made in the management of residuals from process industries in terms of regulations and technical solutions, there is still a long way to go in addressing the material imbalance that exists in these industries.

Therefore, the utilization of natural resources as efficiently as possible is arguably one of the greatest challenges facing our modern world. There is an ever increasing demand for more efficient processes and equipment partly driven by various emerging environmental problems and stringent regulations. In the light of this prevailing circumstance, process engineering must constantly innovate in order to meet the ever increasing demands for better process equipment and techniques. Consequently, there is a growing interest in the development of a new generation of process reactor systems for use in many industrial downstream operations. This is particularly beneficial considering the increasing pressure on natural resources as a result of the expanding global population.

1.2 STATE-OF-THE-ART IN RESIDUALS PROCESSING

Many conventional and emerging phase separation (solid-liquid) techniques have been documented in scientific literature (Abu-Orf et al., 2004; Mahmoud et al., 2010, 2013; Novak, 2001, 2006; Wakeman, 2007). However, new sludge treatment technologies have not evolved as rapidly as in water and wastewater treatment but some significant improvements have been recorded. In most of these techniques, the aim is either to fundamentally alter the physical state of the sludge matrix so that the solid and the liquid phases can be separated easily or apply external force to drive out water from the floc matrix or a combination of both (Mahmoud et al., 2013). The final achievable dry solids content varies widely across different separation techniques. For instance, conventional mechanical separation techniques such as gravitational settling, filtration and centrifugation are capable of delivering a dry solids content of between 15 and 50% (Mahmoud et al., 2013) (Figure 1.1). Despite limited breakthrough in sludge dewatering improvements, phase separation of fine particulate system has remained a fruitful area of research.

In a study commissioned by the United States Environmental Protection Agency (USEPA), a number of promising embryonic sludge pre-treatment and separation techniques such as electrodewatering, ultrasonic, membrane filtration, mechanical freeze-thaw, chemical cell destruction (micro sludge™), biological cell destruction (BIODET™), and enzyme conditioning were reported. In practice, most of these emerging techniques are employed in conditioning the sludge while the conventional technologies are used in the subsequent dewatering operations as illustrated in Figure 1.2 (Hetherington et al., 2006).

1.3 PROBLEM STATEMENT AND HYPOTHESIS

Over the past decade, several empirical and analytical studies of fluid flow and turbulent aggregation processes in many flow units with different energy dissipation

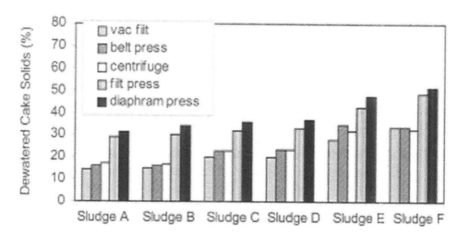

Figure 1.1 Dry solids concentration obtainable with conventional dewatering technologies (Reprinted from Novak, 2001 with permission © 2001 IWA Publishing).

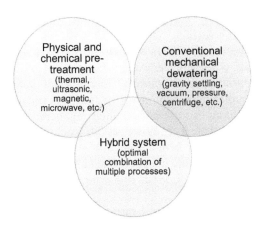

Figure 1.2 Conceptual model of process integration for improved sludge dewatering.

and mass transfer rates have been conducted (Carissimi & Rubio, 2015; Coufort, Bouyer, & Liné, 2005; Sievers et al., 2008; Watanabe & Tanaka, 1999). Most of these studies were largely confined to the use of existing pre-treatment facilities such as annular-based devices or sludge pelletizers with well-defined geometry and stirrer configurations. However, these devices are of limited technical and industrial applications due to their low throughput (i.e. sludge volume treated compared to the reactor size) and process performance.

There is a lack of research work on unconventional or alternative stirrer-vessel systems, partly due to the extreme complexities of the flow structure in such devices (Visscher et al., 2013). Furthermore, limited progress in this area has resulted in slow adoption of many of these devices on industrial scales despite their huge potential.

On the basis of this identified research problem, an innovative design concept was developed that aims to address the problem of throughput constraints and high operating cost in the technical application of classical Taylor-Couette devices. The main objective of this study is to demonstrate the structure formation process in wet agglomeration technique (pelleting flocculation) by simulating the design principle in a rotor-stator cavity as a technical proof of concept with a view to transfer the knowledge into practical application. This is achieved by investigating the process performance using flow units of variable geometries but with identical stirrer configuration, under different operating conditions. A number of process conditions influencing the pre-treatment method were optimized using quasi-factorial experimental design followed by an evaluation of the process performance.

1.4 OVERVIEW AND SCOPE OF STUDY

This study aims to demonstrate that the performance improvement of the agglomeration process can be explained in purely quantitative terms by defining the process performance criteria and specific indicators of process efficiency. A simplified process design approach was adopted to demonstrate the feasibility of the design concept as shown in the flow chart in Figure 1.3. The simplified approach will allow rapid simulation and refinement of the process conditions and test apparatus within the time frame of this research work.

A working hypothesis adopted in this study is based on the premise that the performance of any pre-treatment device is a function of the physicochemical, process engineering or operating conditions and stirrer-vessel system (*configuration and geometry*). Previous studies have demonstrated that it is possible to optimize all the influencing parameters in an evolutionary manner by incremental adjustments of the process conditions (Hessel, Hardt, & Löwe, 2004; Kockmann, 2008).

The implementation of this bench-scale study involves a number of key tasks which were carried out sequentially in order to achieve the aim of the study. The main tasks carried out in this study include the following:

- Comprehensive review of existing literature
- Conceptualization of the processing technique
- Definition of process performance criteria
- Design and fabrication of the flow apparatus
- Optimization of the physicochemical process
- Experimental simulation of the pelleting process
- Modification and refinement of flow apparatus
- Evaluation of process efficiency

1.5   OUTLINE AND STRUCTURE OF THE THESIS

This thesis documents the main findings from this investigation and it is structured into eight chapters. The first chapter provides a general background for this investigation and frames the problem statement. Chapters 2, and 3 provide the theoretical framework for the study and briefly explore the fundamental concepts underpinning the research work. A brief discussion of a number of concepts such as

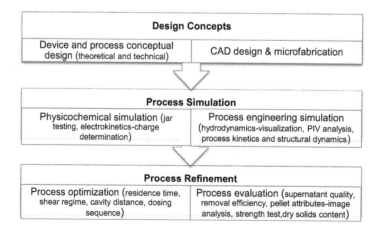

Figure 1.3 Flowchart showing the different aspects of the study.

flocculation theories, structure formation processes, colloidal stability, hydrodynamics interactions and aggregate stability is provided from published scientific literature. Classical and extended Derjaguin-Landau-Verwey-Overbeek (DLVO) theory, kinetic theory of flocculation and sludge pre-treatment techniques are some of the additional concepts briefly reviewed.

In chapter 4, a description of the test materials and experimental methods are presented. This comprises of the characteristics of the test substrates as well as the flow devices used for the experimental simulation. A detailed description of the experimental and analytical methods for data collection and analysis was also provided in this chapter. The design concept is introduced in chapter 5 with detailed description of the stirrer-vessel system followed by theoretical and experimental analysis of the flow streamlines and vortex pattern in the pre-treatment devices. In chapter 6, a theoretical analysis of the mixing and hydrodynamic characteristics is provided.

The optimization of the physicochemical parameters prior to slurry pelletization in agglomeration units along with the operating conditions influencing the pelleting process were discussed in chapter 7. Lastly, chapter 8 provides a summary of this study, highlighting the important findings from the study, and identifying future research needs.

Chapter 2

Fundamentals of flocculation and colloidal stability

2.1 STRUCTURE FORMATION IN DISPERSED SYSTEMS

In environmental management, the removal of particulate solids from liquid process effluents is of great importance. However, when the sizes of solid particles diminish and reach micron and submicron range, the particles tend to remain in suspension and cannot be removed by gravity settling. In order to achieve an acceptable solid-liquid separation at a reasonable cost, the particles need to be agglomerated by flocculation followed by mechanical phase separation (sedimentation, filtration, centrifugation etc.) (Pietsch, 2002). The agglomeration of suspended particles to form larger and settleable flocs form the basis of operation of many process industries ranging from water and wastewater treatment to pulp and paper processing.

Flocculation plays a key role in several natural processes in the aquatic and marine environment such as sediment erosion, transport, aggregation and deposition and it is often employed in several industrial purification and separation processes (Addai-Mensah & Prestidge, 2005; Farinato, Huang, & Hawkins, 1993; Lick, 2008; Lick, Huang, & Jepsen, 1993; Prat & Ducoste, 2007; Prat & Ducoste, 2006; Runkana, Somasundaran, & Kapur, 2006). Aggregation of cohesive suspended particles by size enlargement to form larger flocs is typically encountered in natural systems such as rivers, reservoirs, lakes and estuarine and in engineered systems such as bioreactors and agitation vessels (Curran & Black, 2005; Sievers et al., 2008; Taboada-Serrano et al., 2005; Tooby, Wick, & Isaacs, 1977; Wu, 2008). It is widely employed in both upstream and downstream solid-liquid separation processes where the value of the individual phases are enhanced by destabilization and aggregation of the charged particles in suspension using high molecular weight synthetic polymers (Biggs, 2006; Gregory & Guibai, 1991; Hjorth & Christensen, 2008; Lee et al., 2012; Yukselen & Gregory, 2004).

Typical mean shear rates in these diverse natural and engineered systems range from 2.5 s^{-1} in natural systems to 5000 s^{-1} especially in high shear bioreactors and mixing chambers (Logan, 2012; Milligan & Hill, 1998). A number of interfacial forces and interactions that play key roles in flocculation process are well-understood in the light of classical and extended DLVO theory (Gregory, 1989, 1993; Popa, Papastavrou, & Borkovec, 2010; Bratby, 2006; Bache & Gregory, 2007), and therefore it provides a basis for any theoretical analysis of particle dispersion and colloidal stability.

2.2   COLLOIDAL STABILITY AND INTERFACIAL FORCES

Most colloidal particles in suspension (0.001 – 10 μm) tend to remain suspended and settle very slowly due to the presence of surface charge arising from isomorphic substitution, chemical reaction at the interface or preferential adsorption of ions on the particle surface from the surrounding medium (Benjamin & Lawler, 2013; Partheniades, 2009). The charge on the particle surface is surrounded by excess of oppositely charged ions (counterions) in solution in order to maintain electrical neutrality. The combined interactions of this system of opposite charge cloud is known as the electric double layer (Gregory, 2006b). The developed electrostatic repulsive forces prevent particle aggregation in any colloidal system and hence contribute to their stability in the dispersed system (Shammas, 2005). The diffuse layer and the Stern layer of the charge cloud surrounding a particle together constitute the electric double layer (Marshall & Li, 2014).

A quantitative description of the colloidal stability and aggregation process was facilitated by the assumptions of the DLVO theory which considered interactions between particles as additive (Gregory, 2006b). The DLVO theory was primarily based on two types of interactions between two compact spherical particles—electrostatic (including *Born repulsion*) (Lebovka, 2013), repulsive force due to the electrical double layer and Van der Waals attractive force of the dispersion.

The combination of these two interactions defines a total energy barrier V_T or the so-called *Gibbs interaction energy* as shown in the energy profile in Figure 2.1 (Nopens, 2005; Gregory, 2006b; Moody & Norman, 2005; Laskowski & Pugh, 1992;

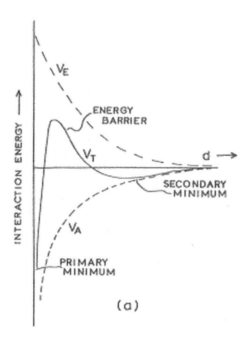

Figure 2.1 Potential energy barrier in the interaction between two particles (Reproduced from Gregory, 1989 © 1989 Taylor & Francis).

Lu, Ding, & Guo, 1998; Bache & Gregory, 2007) which, if overcome by the kinetic energy of the moving particles, will result in aggregation. This is expressed mathematically in Eqs. 2.1-2.2, where V_A, V_R, ζ refers to the energy barrier due to Van der Waals attraction and electrostatic repulsion and zeta potential respectively. In the case of submicron particles, the kinetic energy is derived primarily from Brownian diffusion and less from hydrodynamic (fluid shear) or gravitational forces (differential settling). However, for larger or coarser particles, hydrodynamic interactions such as turbulence or fluid shear and gravitational forces imparts much of the kinetic energy needed to overcome the flocculation barrier (Wilkinson & Reinhardt, 2005; Moody & Norman, 2005; Bagster, 1993). A critical shear rate is required to overcome the energy barrier in order to allow flocculation and floc formation, and this is dependent on the surface charge and the size of particles. In general, the higher the charge on the particles, and the smaller the particle size, the higher the shear rates that are required to bring about flocculation (Gregory, 1989; Lu et al., 1998; Smith-Palmer & Pelton, 2006).

$$V_T = V_R + V_A \qquad (2.1)$$

$$V_T = \pi \varepsilon d \zeta^2 \exp(-kh) - \frac{Ad}{24h} \qquad (2.2)$$

In addition to the interactions described above, there exist additional interaction energies and structural forces which may arise for instance from the perturbation or re-arrangement of water molecules near the interface (Addai-Mensah & Prestidge, 2005; Laskowski & Pugh, 1992; Schramm, 2005). Non-DLVO interactions such as hydrophobic, hydration, oscillatory, structural, steric and electrosteric, and hydrodynamic forces, which were previously not considered in colloidal interactions are now widely recognized to play important roles in colloidal aggregation (Addai-Mensah & Prestidge, 2005; Grasso et al., 2002; Gregory, 1989, 1992, 1993). These forces may be attractive, repulsive or oscillatory in nature and may be more pronounced or stronger than the DLVO forces (Addai-Mensah & Prestidge, 2005). In some cases where the energy barrier for aggregation exists at subnanometer distances, significant discrepancies between theoretical (DLVO-based) and experimental measurements of flocculation rates have been reported (Taboada-Serrano et al., 2005; Wilkinson & Reinhardt, 2005). For instance, in the case of colloidal dispersion containing clay particles, the theory provides a general conceptual model for the gross interactions of particles but fails to explain certain aspects of the interactions (Partheniades, 2009). However, in most practical applications of flocculation, the dominant forces are electrostatic repulsion and London-Van der Waals attraction forces (Hanson & Cleasby, 1990).

2.3   KINETICS OF FINE PARTICLE AGGREGATION

Flocculation kinetics deal with time-dependent changes in the dispersion or suspension and provides information on the flocculation rate, dispersion stability and particle interactions which depends on the number and efficiency of the particle collisions

(Kissa, 1999). The collisions between particles has been suggested to occur at a rate that depends on the transport mechanisms of Brownian diffusion, fluid shear and differential settling and it is assumed to be a second-order rate or reaction process (Gregory, 2013a). The aggregation rate of dilute systems (TS ≤ 1%) with an initial concentration of primary particles n_1 and fast aggregation constant k^f can be represented by the expression in Eq. 2.3 (Gregory, 2013a; Lebovka, 2013).

$$\frac{dn_1}{dt} = -k^f n_1{}^2 \tag{2.3}$$

Furthermore, several studies have shown that aggregates formed from different transport mechanisms exhibit marked variations in their structural attributes. The aggregates formed by shear-induced collision are known to be stronger than those formed from other transport mechanisms, those from Brownian motion tend to be easily dispersed by shearing, while those from differential settling results in ragged, weak, and low density aggregates (Van Leussen, 2011).

Smoluchowski using the "population balance" approach, expressed time evolution of the number density of discrete particles of size k as they aggregate with respect to time in terms of the collision frequency or rate function β originally for Brownian diffusion and laminar or uniform shear flow, and later for differential settling assuming binary collisions between the particles (Atkinson, Chakraborti, & Benschoten, 2005; Gregory, 1993, 2013a; Kramer & Clark, 1999; Thomas, Judd, & Fawcett, 1999). Although the Smoluchowski equation has been modified in line with recent findings to include several other parameters, the general form of the expression is presented in Eq. 2.4. The quantity β_{ij} is the collision frequency or rate function for collisions between ith and jth sized particles while α_{ij} is the dimensionless collision efficiency factor. The collision frequency depends on the transport mechanisms of Brownian motion, fluid shear and differential settling, while the collision efficiency is a function of the degree of particle destabilization and it gives the probability of collision leading to attachment with values ranging from 0 to 1 (Kruster, 1991; Lawler, 1993; Lick, Lick, & Ziegler, 1992a, 1992b; Thomas et al., 1999).

$$\frac{dn_k}{dt} = \frac{1}{2} \sum_{i=1,\, i+j=k}^{k-1} \alpha_{ij}\beta_{ij}\left(i,j\right)n_i n_j - \sum_{i=1}^{\infty} \alpha_{ik}\beta_{ik}\left(i,k\right)n_i n_k \tag{2.4}$$

The first term of the right hand side of Eq. 2.4 describes the increase in number of particles of size k by flocculation of two particles whose total volume is equal to the volume of a particle of size k, while the second term on the right hand side describes the loss of particles of size k due to aggregation with particles of other sizes. The general form of the equation expresses the rate of change in the number concentration of particles of size k. In arriving at Eq. 2.4, Smoluchowski made a number of key assumptions listed below:

•   The collision efficiency factor α is unity for all collisions
•   The fluid motion undergoes laminar shear
•   The particles are mono dispersed

- There is no breakage of flocs
- All particles are spherical in shape before and after collision
- Collision involves only two particles

The interparticle collisions are promoted by transport mechanisms of Brownian motion or perikinetic (for particles with diameters less than 1 µm), fluid shear or orthokinetic (for particles in the diameter range 1-40 µm) and differential settling (typically for particles with diameter larger than 40 µm). The collision frequency or rate function β for these three mechanisms can be expressed mathematically in Eqs. 2.5-2.7 where β_{BM}, $\beta_{SH,}$ β_{DS} refers to the collision frequency function for Brownian motion, fluid shear, and differential settling respectively. The total collision rate or frequency β_T is the sum of contributions from each of the three transport mechanisms (Eq. 2.8) (Atkinson et al., 2005; Lick, 2008; Lick et al., 1992b, 1993; Lick & Lick, 1988; Tsai, Iacobellis, & Lick, 1987; Van Leussen, 2011; Wang et al., 2005; Wu, 2008).

$$\beta_{BM} = \frac{2kTd_{ij}^{2}}{3\mu d_i d_j} \tag{2.5}$$

$$\beta_{SH} = \frac{G(d_i + d_j)^3}{6} \tag{2.6}$$

$$\beta_{DS} = \frac{\pi g}{72 v \rho_w} (d_i + d_j)^2 \ (\Delta \rho_i d_i^{2} - \Delta \rho_j d_j^{2}) \tag{2.7}$$

$$\beta_T = \beta_{BM} + \beta_{SH} + \beta_{DS} \tag{2.8}$$

2.3.1 Perikinetic particle aggregation

The random displacement of particles in Brownian motion as a result of the thermal energy of the system is termed perikinetic (Gregory, 1989, 1992, 1993, 2013a). In practical applications, perikinetic aggregation is only important for very small particles where they are continuously bombarded by the surrounding water molecules (< 1 µm) (Gregory, 2006a). Under this prevailing condition, the rate of floc growth cannot be sufficiently sustained for an effective phase separation. Hence, the importance of other transport mechanisms in promoting the floc growth kinetics. However, in the case of contact between a small particle and micro flocs, Brownian diffusion can still control the transport of small particles across the layer of fluid on the floc surface (Letterman, Amirtharajah, & O'Meila, 2010).

2.3.2 Hydrodynamic-mediated interactions

2.3.2.1 Classical orthokinetic aggregation

The agitation of suspended particles in liquid medium will lead to aggregation due to the inter-particle collision induced by the fluid motion (Bagster, 1993). The velocity gradient or shear rate is dependent on the nature of the fluid flow. In uniform laminar shear flow, the velocity gradient remains constant in the entire flow field while in turbulent flow, there is a rapid fluctuation of the velocity gradient. Consequently, the velocity gradient is a function of both space and time (*spatial and temporal*) (Bridgeman, Jefferson, & Parsons, 2008, 2010). Camp and Stein in their critique of the Smoluchowski approach to shear flocculation modelling (Camp & Stein, 1943), introduced the concept of root-mean-square (R.M.S.) or absolute velocity gradient $\bar{G}$ to account for the variations in the shear rate. (Benjamin & Lawler, 2013; Letterman et al., 2010; Van Leussen, 2011; Winterwerp, 1998; Zhu, 2014). Considering the angular distortion of an elemental volume of fluid arising from the application of tangential surface forces, $\bar{G}$ is defined according to Eqs. 2.9-2.10 as the R.M.S. velocity gradient in the mixing vessel, where u, v, w are the components of the fluctuating velocity while x, y, z refers to the 3-D Cartesian coordinate system, P is the power dissipated, V is the volume of the mixing vessel, ε is the energy dissipation rate per unit mass, μ is the dynamic viscosity and v is the kinematic viscosity (Benjamin & Lawler, 2013; Bridgeman, Jefferson, & Parsons, 2009).

$$\bar{G} = \left\{ \left(\frac{\partial u}{\partial y} + \frac{\partial v}{\partial x} \right)^2 \left(\frac{\partial u}{\partial z} + \frac{\partial w}{\partial z} \right)^2 \left(\frac{\partial v}{\partial z} + \frac{\partial w}{\partial y} \right)^2 \right\}^{\frac{1}{2}} \tag{2.9}$$

$$\bar{G} = \sqrt{\frac{P}{\mu V}} = \sqrt{\frac{\varepsilon}{v}} \tag{2.10}$$

In practical applications, owing to the difficulties associated with the calculation of absolute velocity gradient $\bar{G}$ due to the fluctuations in the energy dissipation within the mixing vessel, an average velocity gradient G_{ave} has often been used in place of the absolute value $\bar{G}$ (Bridgeman et al., 2008; Bridgeman et al., 2010) (Eq. 2.11). Korpijarvi et al. (Korpijärvi, Laine, & Ahlstedt, 2000), in their study of mixing and flocculation in a jar using computational fluid dynamics (CFD) reported a large variation in the local velocity gradient G_L within the mixing vessel. In similar studies conducted by Kramer and Clark (1997), Benjamin and Lawler (2013), as well as Mühle (1993), a lower estimate of the absolute velocity gradient was proposed (Eqs. 2.12-2.13) to account for the variance of the velocity gradient throughout the fluid where quantities G_L, P_{ave}, ε, V, μ, v represent the local velocity gradient at different points within the mixing vessel, average power consumption, kinetic energy dissipation rate, volume of the mixing vessel, dynamic and kinematic viscosity respectively.

$$G_{ave} = \sqrt{\frac{P_{ave}}{\mu V}} = \sqrt{\frac{\varepsilon_{ave}}{v}} \tag{2.11}$$

$$\bar{G} = \sqrt{\frac{P}{\mu V} - \sigma_{GL}^2} \tag{2.12}$$

$$\bar{G} = 0.26 \sqrt{\frac{P}{V\mu}} = 0.26 \sqrt{\frac{\varepsilon}{v}} \tag{2.13}$$

2.3.2.1.1 Particles in laminar shear

According to the rectilinear model of particle trajectory, two particles moving in a fluid on different streamlines will experience a velocity gradient which indicates the relative motion of particles and the possibility of collision and aggregation as illustrated in Figure 2.2 (Gregory, 2013a; Svarovsky, 2000). In simple laminar or uniform shear with a well-defined flow field such as in pipe flow or in some Couette devices (Belfort, 1986; Ives, 1984; Logan, 2012), the transport of particles by laminar shear due to the fluid motion can be characterized by a single value of the shear rate G (Benjamin & Lawler, 2013). Under such condition, the particles in laminar shear will exhibit rotational motion in the direction of travel with a constant angular velocity ω (Falk & Commenge, 2009; Spicer, 1997).

2.3.2.1.2 Particles in turbulent shear

In most practical applications of flocculation, the aggregation processes typically occur under turbulent conditions (Benjamin & Lawler, 2013; Carissimi & Rubio, 2015; Concha, 2014; Farrow & Swift, 1996; Svarovsky, 2000). The particles in suspension under the influence of turbulent eddies will experience fluctuating motion of the fluid with the particles being transported by the fluid eddies and the flow vortex (Benjamin & Lawler, 2013; Letterman et al., 2010). Thus, small particles suspended in fluid exist in an environment of small energy-dissipating eddies in most typical sheared reactors (Hendricks, 2011; Logan, 2012). Under such condition, particle collisions are promoted by eddy size similar to that of the colliding particles as shown in Figure 2.3 (Shamlou & Hooker-Titchener, 1993; Thomas et al., 1999). The mechanism of flocculation in turbulent shear has been suggested to be similar to that of laminar shear for particles smaller than the Kolmogorov microscale λ_0 of turbulence, while for particle larger than this length scale, the flocculation mechanism is similar to that of Brownian diffusion (Benjamin & Lawler, 2013; Gregory, 1989).

The formation of aggregates in any turbulence or shear-induced flocculation essentially consist of the following phases: destabilization; collision and adhesion; floc formation, growth and deformation phases (Hogg, 2000). Ives (1984) and Bergenstahl (1995) expressed the kinetics of orthokinetic aggregation in terms of

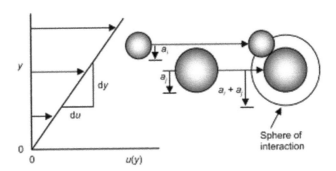

Figure 2.2 Conceptual view of particle transport in simple laminar shear leading to collisions (Reproduced from Benjamin & Lawler, 2013 with kind permission © 2013 Wiley).

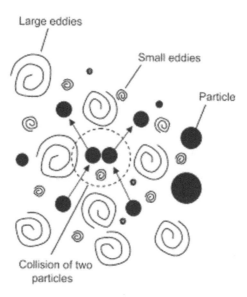

Figure 2.3 Schematic illustration of particle—particle and eddy—particle interactions in turbulent flow (Reprinted from Yeoh, Cheung, & Tu, 2014 with permission © 2014 Elsevier).

the change in the initial particle concentration, while Levich (Son & Hsu, 2008; Winterwerp, 1998) in an earlier study derived the collision rate equation based on the concept of locally isotropic turbulence assuming a viscous diffusive subrange (i.e. particle size smaller than Kolmogorov micro scale) for shear-induced turbulent flow where N, r, d represent the particle number concentration per unit volume, particle radius and diameter respectively (Eqs. 2.14-2.15). Several other models of flocculation kinetics are available in literature (Adachi, Kobayashi, & Kobayashi, 2012; Partheniades, 2009). In most of these models, the shear rate G and the kinetic energy dissipation rate ε are shown as important drivers of the flocculation process. Typically, orthokinetic flocculation can be experimentally observed in a conventional stirred tank consisting of an axially mounted impeller in a circular or rectangular mixing vessel.

$$\frac{-dN}{dt_{(Couette)}} = \frac{2}{3}f_o Gd^3 N^2 + \frac{4}{3}f_p \frac{kTN^2}{\mu} \tag{2.14}$$

$$\frac{dN}{dt_{Turb.}} = -6\pi\beta\sqrt{\frac{\varepsilon_{intensity}}{\eta}}\, r^3 N^2 = -12\pi\beta\sqrt{\frac{\varepsilon_{intensity}}{\eta}}\, d^3 N^2 \tag{2.15}$$

2.3.3   Extended orthokinetic flocculation

Pelleting flocculation is considered as an extension of the classical orthokinetic transport mechanism based on the "metastable state" concept (Norihito Tambo,

1990; Yusa, Suzuki, & Tanaka, 1975). It has been shown by several studies that the efficiency of the floc formation process as well as the floc structural attributes (size, shape, porosity, density, etc.) can be significantly improved by the application of suitable mechanical energy (Bähr, 2006; Gang et al., 2010; Gillberg et al., 2003; Glasgow, 2005; Higashitani, Shibata, & Matsuno, 1987; Panswad & Polwanich, 1998; Walaszek, 2007; Wang et al., 2004; Yusa & Gaudin, 1964; Yusa & Igarashi, 1984; Yusa, 1977). This is accomplished by the application of a non-destructive uneven force system on the floc surface by the action of the turbulent fluid motion or the so-called floccule mechanical synaeresis induced by rolling and collision mechanisms (Amirtharajah & Tambo, 1991; Hemme, Polte, & Ay, 1995; Higashitani & Kubota, 1987; Vigdergauz & Gol'berg, 2012; Yusa, 1987).

In the rolling mechanism, as the floc rolls along a plane surface, it is subjected to a pressure fluctuation which becomes greater at the front of the central point in the contact area and lesser at the rear. On the other hand, in the collision mechanisms, a floc experiences an impact due to the floc-floc or floc-surface collision. In order to withstand this impact, the floc must adhere strongly, in order words, the impact results in a more compact floc (Yusa et al., 1975). In practice, pelleting flocculation is realized through the selection of an appropriate process engineering conditions and stirrer-vessel system (geometry and configuration) in order to obtain the necessary conditions suitable for rolling and collision-mediated floc pelletization (Walaszek, 2007; Yusa et al., 1975).

2.3.3.1 Mechanisms of pelleting flocculation

The structure formation in pelleting flocculation is an extension of the classical orthokinetic flocculation facilitated by the application of suitable external forces. The flocs obtained from the process referred to as "pellet flocs" exhibit superior structural attributes when compared to the "random flocs" produced by the classical flocculation process which tend to be loose and bulky (Yusa & Igarashi, 1984). Previous studies have reported two distinct models of pelleting flocculation, namely parallel and series models resulting in onion and raspberry-like pellet floc structures as presented schematically in Figure 2.4 (Bähr, 2006; Tambo & Wang, 1993; Walaszek, 2007; Walaszek & Ay, 2005, 2006a, 2006b; Yusa, 1977). In the series model, orthokinetic flocculation occurs first where random, loose and bulky flocs are produced. These voluminous flocs shrink and densify by the application of suitable mechanical energy (mechanical synaeresis).

By contrast, in the parallel model, there is no time delay between the orthokinetic flocculation and mechanical synaeresis. The dispersed particles in suspension and micro flocs are simultaneously transported by the fluid motion to the surface of the "mother seed" where they are attached by polymer bridging (Walaszek & Ay, 2005, 2006a; Yusa, 1987). In terms of the flocculation efficiency and process relevant parameters, pellet flocs can be characterized by their size, shape, density, porosity, compressive strength and deformability. A comparison of the relevant properties of conventional and pellet flocs is provided in Table 2.1.

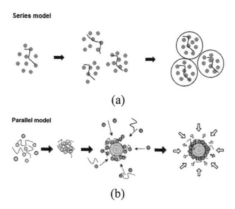

<figure>Figure 2.4 Conceptual models of pelleting flocculation process (a) series model—raspberry structure (b) parallel model—onion structure.</figure>

Table 2.1 A comparison of process relevant characteristics of classical and pellet flocs

Features	Conventional flocs	Pellet flocs
Size	Diameter (up to 2 mm)	Diameter (up to 10-20 mm)
Form	Irregular	Nearly spherical
Density	1.00-1.01 gcm^{-3}	Up to 1.1 gcm^{-3}
Strength	Low	High
Settling	Modest	Fast
Dewaterability	Satisfactory	Good
Deformability	Strongly	Slightly

2.3.3.2 Pelleting reactor systems

2.3.3.2.1 Annular-based reactors

In the annular-based or Taylor-Couette reactors shown schematically in Figure 2.5, the pelleting process takes place within the annular gap of concentric cones or cylinders. The inner cylinder or cone with possibility for axial adjustment is allowed to rotate during the pelleting process while the outer cylinder or cone remains stationary. The propagated Taylor vortices between the gap along with the reactor geometry induces the necessary rolling and collision effects. The flocculant and slurry are fed into the annular gap of the reactor while the rotational flow promotes contacts between the particles and the polymer molecules. Unattached slurry particles are recirculated back into the reactor through an external loop. The main drawback of these annular-based system is the small space available for slurry treatment in comparison to the reactor volume (Hemme et al., 1995; Sievers et al., 2008).

2.3.3.2.2 Sludge blanket clarifiers

The operation of the sludge blanket clarifiers or pelleting flocculation blanket (PFB) shown in Figure 2.6 is based on the appropriate control of the slurry up-flow rate

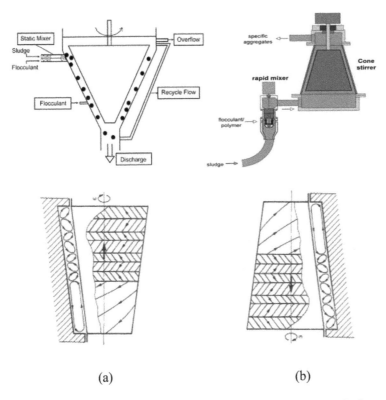

(a)   (b)

Figure 2.5 Schematic diagram of two common types of annular reactors showing the flow streamlines in the annulus (a) conical Couette reactor (Hemme et al., 1995; Wimmer & Zierep, 2000) (b) cone-stirring reactor (Reproduced from Sievers et al., 2008 ; Wimmer, 1995 with kind permission © 1995 Cambridge University Press).

and agitation condition. This promotes one-by-one attachment of primary flocs to the grown floc shell resulting in a compact aggregate. (Gang et al., 2010; Watanabe & Tanaka, 1999). In this process, the first important factor for pellet formation is adequate mixing to promote contact between the sludge particles and the polymer molecules. The second factor is the rolling action required for the floc growth, which is realized by both vertical and horizontal circulating flows along the wall of the tank. The main difference between this system and the annular-based reactors is the fact that the build-up of the pelleting process in sludge clarifiers is non-uniform due to a single inlet point for slurry and flocculants. This consequently reduces the probability of incoming floc attachment to the pelleted "mother flocs" in order to form onion-like pellet flocs as compared to the Taylor-Couette systems (Walaszek & Ay, 2006a).

2.3.3.3 Structure and morphology of pellet flocs

The structure of pellet flocs differs significantly from that of a conventional amorphous floc. They are characterized by their large size, narrow size distribution, high density and nearly spherical shape (Higashitani & Kubota, 1987). The difference stems from

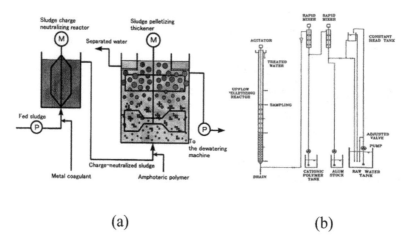

(a)   (b)

Figure 2.6 Schematic diagram of two common types of sludge blanket clarifiers (a) pelleting flocculation blanket (PFB) (Reproduced from Watanabe & Tanaka, 1999 with permission © 1999
Elsevier) (b) fluidized-bed pellet contact clarifier (Reproduced from Panswad & Polwanich,
1998 with permission © 1998 IWA Publishing).

the fundamental differences between classical orthokinetic and pelleting flocculation
processes. Two distinct forms of pellets flocs have been reported in literature: raspberry
structure and onion structure shown in Figure 2.7 (Walaszek & Ay, 2006a).

In the raspberry-like structure obtained from the series model, there is a random
attachment and growth of the agglomerate from the micro particles and micro flocs
to form macro floc followed by the exudation of the dispersion medium. By contrast,
in the onion-like structure obtained from the parallel model, the dispersed particles
and micro flocs are attached to the "pelleted mother floc" and densify simultaneously
thereby forming a layered structure. In practice, the onion-like structure requires an
enhanced degree of process control which is extremely difficult to obtain in full-scale
operations (Walaszek & Ay, 2006a).

Several studies have reported that water exists in different forms in a sludge floc and
the quantity of water that can be removed during the pre-treatment process is dependent
on the status of the water in the floc (Colin & Gazbar, 1995; Gillberg et al., 2003; Lee
& Hsu, 1995; Outwater & Tansel, 1994; Sanin, Clarkson, & Vesilind, 2011; Vesilind,
1994; Vesilind & Tsang, 1990; Wu, Huang, & Lee, 1998). The physical state of water
in a sludge floc has been suggested to exist in four distinct categories. Free or bulk water
exists independently of the agglomerated particles, interstitial water is contained in the
space within the floc, vicinal or surface water is tightly bound to the surface of the particles by hydrogen bonding, while water of hydration or bound water is chemically bound
to the particle surface (Novak, 2006). In a another study, the distribution of moisture
in an ideal spherical particulate network was reported to exist in three distinct forms,
namely capillary, funicular and pendular (Besra, Sengupta, & Roy, 1998). The difficulties in dewatering sludge arise from its colloidal and compressive nature as well as the
presence of organic substances (bacteria cells and extracellular polymeric substances) in
the case of biological sludge. It is therefore necessary, prior to dewatering, to condition

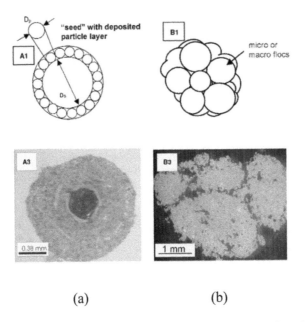

Figure 2.7 Schematic illustration of the structure and morphology of pellet flocs (a) onion structure (b) raspberry structure (Reprinted from Walaszek, 2007 with permission).

sludge using chemicals such as flocculants or a variety of other physical pre-treatment methods mentioned in the previous chapter to alter the structure of the flocs with respect to dewatering kinetics and attainable dry solids content (Mahmoud et al., 2013).

2.3.4   Polymer-mediated interactions

The energy barrier to flocculation can be overcome in several other ways apart from the kinetic energy derived from the transport mechanisms. The electrostatic charge can be lowered with the aid of polymeric flocculants by double layer compression, charge neutralization, and polymer bridging between particles (Hjorth, 2009; Moody & Norman, 2005; Nopens, 2005). In practice, both of these approaches are employed either concurrently or sequentially. In addition to these interactions, several other mechanisms have also been attributed to the polymer flocculation of charged particles in suspension. Electrostatic charge-patch; depletion flocculation; network flocculation and polymer complex formation are few of the other relevant interactions (Addai-Mensah & Prestidge, 2005; Concha, 2014; Farinato et al., 1993; Moudgil, 1986; Petzold & Schwarz, 2013; Smith-Palmer & Pelton, 2006; Wang & Jiang, 2006; Wilkinson & Reinhardt, 2005; Xiao, 2006).

However, the dominant concepts in the case of polyelectrolyte-mediated destabilization and aggregation of charged particles in suspension are surface charge neutralization (ion exchange), charge patch formation and polymer bridging as illustrated in Figure 2.8 (Besra et al., 2002; Böhm & Kulicke, 1997; Hjorth, Christensen, & Christensen, 2008; Hjorth & Jørgensen, 2012; Kissa, 1999; Lagaly, 1993,

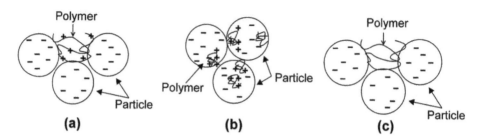

Figure 2.8 A schematic representation of the dominant mechanisms in polymer-mediated interactions (a) charge neutralization (b) charge patch formation (c) polymer bridging (Reproduced from Lagaly, 2005 © 2005 Taylor & Francis).

2005; Lee & Liu, 2000). The important physicochemical properties that influence the flocculation process in the case of polymer-mediated interactions include; polymer molecular weight, charge density and concentration, zeta potential or surface charge of the particles in suspension, particle size and distribution, specific surface area, solution conductivity and pH (Wang & Jiang, 2006).

In the case of pelleting flocculation, Higashitani and Kubota (1987) in their study of the pelleting flocculation of polystyrene latex particles reported that the formation of pellet flocs depends on the suspension concentration, flocculant molecular weight and charge density as well as the intensity of mixing. They argued that pellet flocs are formed within a limited range of these parameters hence pelleting flocculation requires a high degree of control of different process conditions. On the basis of their findings, they expressed mathematically the volume fraction of particles φ_{pt} at the critical particle concentration N_c above which pellet flocs are formed as a function of the initial particle concentration N_0, polymer concentration C_p, charge density p, and the molecular weight M, where N_A is the Avogadro constant (Eq. 2.16).

$$\phi_{pt} = 4 \times 10^{-5} \frac{C_p N_A}{p N_0 M} = 24.08 \times 10^{18} \frac{C_p}{p N_0 M} \qquad (2.16)$$

2.3.4.1   Charge neutralization

The concept of charge neutralization or ion exchange (Figure 2.8a) describes the destabilization of an electrostatically stabilized system by the adsorption of oppositely charged polyelectrolyte molecule to the particle surface thereby lowering the net surface charge. Under such condition, the maximum flocculation performance will be attained when the net surface charge of the system is close to zero (Smith-Palmer & Pelton, 2006). This type of interaction typically involves ionic polymers of low to medium molecular weight but high charge density (Moudgil, 1986).

2.3.4.2   Polymer bridging

The bridging mechanism (Figure 2.8b) has been suggested to occur as a result of the adsorption of polymer chains to several particles simultaneously, thereby forming molecular bridges between adjoining particles in the floc. Several factors has been described to play important roles in this process such as chemical, electrostatic, and hydrogen bonding, as well as Van der Waals and hydrophobic interactions (Addai-Mensah & Prestidge, 2005). This type of interaction typically involves high molecular weight polymers with a charge density less than 15% (Smith-Palmer & Pelton, 2006).

2.3.4.3 *Charge-patch formation*

Electrostatic charge-patch model of destabilization (Figure 2.8c) has been described to occur in the case of highly charged cationic polyelectrolyte (polymer charge density > particle charge density), which adsorb to the particle surface with flat configuration that is not suitable for bridging. In the presence of particles whose surface are not completely covered by the polymer, the polyelectrolyte molecule adsorbs to form a cationic patch or charge mosaic on an anionic particle surface thereby allowing a negatively charged particle to attach to the positive patch (Popa et al., 2010; Smith-Palmer & Pelton, 2006).

2.4 FLOCCULATION ASSESSMENT TECHNIQUES

In assessing flocculation performance under laboratory conditions, several simulation tests and measurement techniques have found to be useful in applications such as the treatment of suspensions and fine particulate systems (Bache et al., 2003; Bache & Zhao, 2001; Gregory, 2009; Smith-Palmer & Pelton, 2006). For instance, the conventional cylinder or Imhoff cone and jar tests, Buchner-funnel or pressure filtration test, CST-test, electrokinetic charge analysis—zeta and streaming potential, colloid titrations, and non-conventional procedures such as photometric dispersion analysis technique, HNMR spectroscopy and fibre optic sensor technique, have all been successfully employed to assess the flocculation efficiency in bench and full-scale studies (Abu-Orf & Dentel, 1997; Ay, et al., 1992; Bache & Zhao, 2001; Bartelt et al., 1994; Bratby, 2006; Byun et al., 2007; Chakraborti, Atkinson, & Benschoten, 2000; Dentel, Abu-Orf, & Walker, 1998; Dentel, Wehnes, & Abu-Orf, 1994; Dentel & Abu-Orf, 1995; Farrow & Warren, 1993; Gregory & Nelson, 1984; Gregory, 2009, 2013b; Guida et al., 2007; Mikkelsen, 2003; Papavasilopoulos, 1997; Ramphal & Sibiya, 2014; Richter et al., 1995; Tambo, 1990; Tarleton & Wakeman, 2006; Teefy, Farmerie, & Pyles, 2001; Von Homeyer et al., 1999). The assessment of the flocculation efficiency is carried out either by direct in-situ observation such as particle counting or by other indirect indicators such as turbidity, settling rates, specific volume of sediment, etc. as illustrated in Figure 2.9. The "optimum" flocculation condition will roughly correspond to the minimum characteristic values of the observed parameters in a parametric behavior-dose chart (Bache & Gregory, 2007).

Consequently, in the design of separation and purification devices, flocculation process is often tailored to the specific solid-liquid separation method (e.g. floatation, centrifugation, filtration, sedimentation etc.) and the intrinsic properties of the slurry as the floc characteristics necessary for different separation methods are unique

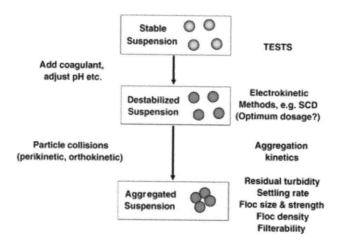

Figure 2.9 Experimental methods for flocculation kinetic measurement (Reprinted from Gregory, 2009 with kind permission © 2009 Elsevier).

(Farinato et al., 1993; Moudgil, 1986). In simulating the physicochemical or flocculation process under laboratory conditions, the aim is to imitate the essentials of full-scale operating conditions as far as possible at a reasonable cost (Dentel, 2010). This is achieved by defining the process performance criteria and appropriate indicators or measurement parameters.

Chapter 3

Hydrodynamics and floc stability in sheared systems

3.1 HYDRODYNAMICS AND FLUID-PARTICLE INTERACTIONS

Hydrodynamically-induced turbulent shear is an important driver of the flocculation process especially in the case of orthokinetic aggregation of particles (Coufort et al., 2005; Moody & Norman, 2005; Wilkinson & Reinhardt, 2005). The floc growth and stability in any flocculated system has been suggested to be a function fluid-particle interactions and the intrinsic physicochemical properties of the floc (Boyle et al., 2005; Carissimi & Rubio, 2015; Shamlou & Hooker-Titchener, 1993). The dynamics of these interactions affects all facets of the flocculation process and the degree of aggregation in any sheared system (Attia, 1992). In the case of hydrodynamic interactions, induced velocity gradient promote the aggregation process but might also be responsible for floc breakage as a result of increased viscous shear stress (Rulyov, 2010). Consequently, in the case of shear-induced collisions, the hydrodynamic effect can be very significant (Gregory, 2006a).

The hydrodynamics and transport process in natural and engineered systems is often controlled more by the fluid properties, in this case the mixing rate of the fluid, than by chemical properties of the particles such as the molecular diffusivity. Therefore, in the design of flocculation reactors, it is desirable to have uniform mixing intensities throughout the reactor (Benjamin & Lawler, 2013). Under the condition of shear-induced turbulence due to the use of flow inducers such as paddles, mixers, stirrers or bubbles, the convective currents in the system are insignificant as compared to the advective currents generated in the system (Logan, 2012). The analysis and modelling of the phenomena of fluid-particle mixing, floc formation, growth and breakup is very complex, as such minimizing the energy dissipation rate (degree of turbulent mixing) in the reactor will lead to improvement in flocculation efficiency (Benjamin & Lawler, 2013).

3.1.1 Fluid mixing and particle dispersion

Mixing can be described as a random distribution of two or more separate phases through one another through the transport mechanisms of advection, turbulence and diffusion occurring at different scales as illustrated schematically in Figure 3.1.

It is typically performed in the process industries and in many engineering applications using a number of techniques, ranging from simple dispersion, blending,

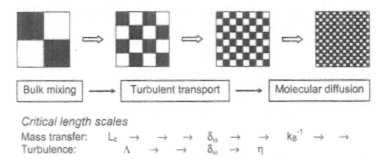

Figure 3.1 Schematic illustration of the transport mechanisms in fluid mixing (Reproduced with from Bache & Gregory, 2007 with kind permission © 2007 IWA Publishing).

agitation or sparging to static flow manipulation (Edwards & Baker, 1997; Hendricks, 2011; Marshall & Bakker, 2004). Advective transport mechanism refers to the movement of a fluid mass where the inertia of the flow results in large eddy formation. Turbulence on the other hand generate velocity gradient or fluid shear which results in the formation of eddies and flow vortex (Hendricks, 2011). Molecular diffusion is ultimately the only process able to mix components of a fluid on the molecular scale (Falk & Commenge, 2009).

3.1.2 Micro, meso, and macro mixing

Turbulent mixing characterized by high Reynolds number ($R_e > 10^4$) (Zlokarnik, 2008), can be viewed as a hierarchy of irregular, rotational and dissipative motion containing vorticity (curl or rotation of the velocity vector) on different scales or eddy sizes (Bache & Gregory, 2007; Baldyga & Bourne, 1984). In turbulent flows such as those encountered in pipe or tube flow, channel flow and stirred tanks or mixing chambers, energy transfer occurs on different eddy scales (Bridgeman et al., 2009; Mühle, 1993). Eddies are spatially recognizable flow patterns that exist in a turbulent flow for at least a short time in which there is a correlation between the velocities at two different points (Benjamin & Lawler, 2013). Fluid deformation causes vortices to stretch and vorticity and kinetic energy to be transported from larger to smaller eddies (Bache & Gregory, 2007; Baldyga & Bourne, 1984). The turbulent vortex in such fluid motion is generally propagated in the tangential and axial direction than in the radial direction with the vorticity increasing with decreasing eddy size (Baldyga & Bourne, 1984; Bridgeman et al., 2009; Thomas et al., 2015).

In most practical applications of mixing, there exist three scales of mixing namely: macro, meso and micro mixing. Macromixing refers to mixing that is driven by the largest scale of motion in the fluid Λ (integral length scale), meso mixing on the other hand involves mixing on a scale smaller than the bulk circulation but larger than the micro mixing, while micro mixing is the mixing on the smallest scale of fluid motion λ_0 (Kolmogorov microscale) expressed in Eq. 3.1 and at the final scales of molecular diffusivity (Batchelor scale) (Bałdyga & Pohorecki, 1995; Kresta & Brodkey, 2004; Sparks, 1996; Thoenes, 1998; Zlokarnik, 2008). Micromixing describes the process homogenization of liquid balls with their surroundings on a molecular level. In typical mixing conditions, the dividing line between micro and macro scale is between 100 and 1000 μm (Oldshue & Trussell, 1991).

$$\lambda_0 = \left[\frac{v^3}{\varepsilon}\right]^{\frac{1}{4}} \tag{3.1}$$

The largest eddies in turbulent dispersion that represent the macro scale of turbulence or the integral length scale (Λ) contains most of the energy, and are produced by the stirrer or agitator head. The size of this macroscale of turbulence is of the order of the diameter of the agitator (Kruster, 1991; Logan, 2012; Zlokarnik, 2008). The turbulent flow can be viewed as an eddy continuum, with their size ranging from the dimension of the turbulence generating device to the Kolmogorov length scale (Kruster, 1991). In between the energy-containing eddies of the integral length scale Λ at the upper end of the inertia subrange and the smallest eddies of Kolmogorov microscale λ_0 at the beginning of viscous dissipation range (Kockmann, 2008), there exist many eddies of other scales smaller than the integral scale Λ that transfer kinetic energy continually through the other length scales. The Batchelor λ_B and Taylor λ_T scales expressed in Eqs. 3.2-3.3 are the examples of other important length scales where D is the diameter of the flocs while ε is the turbulent kinetic energy dissipation rate.

$$\lambda_B = \left[\frac{D^2 v}{\varepsilon}\right]^{\frac{1}{4}} \tag{3.2}$$

$$\lambda_T = \frac{u\sqrt{15}}{\sqrt{\frac{\varepsilon}{v}}} \tag{3.3}$$

The Taylor scale λ_T is an intermediate length scale in the viscous subrange that is representative of the energy transfer from large to small scales, but it is not a dissipation scale and does not represent any distinct group of eddies (Maggi, 2005). Batchelor micro-scale on the other hand is a limiting length scale where the rate of molecular diffusion is equal to the rate of dissipation of turbulent kinetic energy and represent the size of the region within which a molecule moves due to diffusional forces (Baldyga & Bourne, 1999; Kockmann, 2008; Kresta & Brodkey, 2004; Sparks, 1996).

3.1.3   Characterization of fluid-particle mixing

Fluid flow and fluid-particle mixing is typically studied in the laboratory using a number of visualization methods (digital image video recordings), tomography, and other advanced laser-based diagnostic techniques such as Particle Image Velocimetry (PIV), Particle Tracking Velocimetry (PTV), Laser Doppler Velocimetry (LDV), Laser Doppler Anemometry (LDA), and Laser-Induced Fluorescence (LIF)

(Kresta & Brodkey, 2004). Particle Image Velocimetry (PIV) is an established non-intrusive optical flow characterization technique (Brown, Jones, & Middleton, 2004; Mavros, 2001; Myers, Ward, & Baker, 2014; Raffel et al., 2007).

This technique can be used to describe the flow in both qualitative and quantitative terms. PIV enables the time averaged 2-D (two-velocity components) and 3-D (three-velocity components) velocity vectors and vorticities to be measured and displayed graphically by seeding the flow with particles that mimic the motion of the fluid within the observation plane. The simulant fluid in most of the PIV measurements consists of distilled water, glucose or corn syrup, glycerol, silicone oils and CMC solution (sodium carboxymethylcellulose) (Brown et al., 2004). In carrying out PIV analysis, the image of the fluorescent particles are digitally captured and post-processed to obtain a map of the flow field and other hydrodynamic parameters such as the velocity profile and vorticity map (Bugay, Escudié, & Liné, 2002; Coufort et al., 2005; Pechlivanidis, Keramaris, & Pechlivanidis, 2014; Perrard et al., 2000; Thomas et al., 2015).

3.2 TURBULENT AGGREGATION PROCESSES

Floc stability under the influence of hydrodynamic force has been suggested to be a function of floc binding or cohesive force F_B, and hydrodynamic breaking force F_H (Adachi et al., 2012; Bouyer et al., 2005; Bouyer, Liné, & Do-Quang, 2004; Bridgeman et al., 2009; Coufort et al., 2005; He et al., 2015; Kobayashi, Adachi, & Ooi, 1999). While the binding force is determined by the floc structure and physicochemical attributes, flow turbulence is the principal factor in the case of hydrodynamic force (Boyle et al., 2005; Bridgeman et al., 2010; Carissimi & Rubio, 2015). Of these two governing factors of floc stability, turbulence is the least understood owing to its complex nature (Argyropoulos & Markatos, 2015; Gregory, 2013a). Therefore, a detailed analysis of floc stability under turbulent conditions often encountered in natural and agitated systems is not only difficult but often time consuming.

Recent advances in computational fluid dynamics has vastly improved the understanding of turbulence phenomenon. However, numerical modelling of complex turbulent fluid-particle interactions remains a challenge in fluid mechanics and it is computationally intensive (Thomas et al., 2015). In addition, the assumption of the rectilinear motion of particles prior to collision in flocculation modelling has been supplanted by recent findings of curvilinear particle trajectory prior to collisions (Thomas et al., 1999; Zhu, 2014). Particles have been shown to move around other particles in a curvilinear path thereby reducing the probability of collision and attachment (Gregory, 2006a; Thomas et al., 1999).

3.2.1 Energy dissipation in turbulent flow

The power input into any agitated system will eventually dissipates as heat due to viscous forces (Hendricks, 2011). The larger eddies of the order of the agitator diameter

$(\Lambda \sim d_A)$ draw energy from the fluid motion, while the smaller eddies transfer that energy gradually and continually to the smallest eddy, where the energy is ultimately dissipated into heat by friction (Partheniades, 2009). The turbulence parameters of interest in any agitated micro-system (with respect to the mixer capacity and performance) are the Kolmogorov micro scale λ_0, turbulence kinetic energy dissipation rate ε, and agitator tip velocity V_{tip} (Thoenes, 1998; Thomas et al., 2015; Wu & Patterson, 1989).

The Kolmogorov length scale λ_0 is an indicator of the rate of micro mixing and mode of agglomerate deformation due to turbulent shear while the energy dissipation rate ε is the rate of the dissipation of kinetic energy. The tip velocity V_{tip} is a measure of the tangential velocity imparted by the flow inducer, an indicator of the strength of the vortex generated by the flow. Peripheral velocity V_{pkol} on the other hand gives an indication of the velocity of the flow vortex at Kolmogorov micro-scale (Hendricks, 2011; Thoenes, 1998; Thomas et al., 2015; Wu & Patterson, 1989). In typical laboratory mixing experiments (D < 1m), the micro-scale of turbulence predominates. Under such conditions, particle collision is promoted by eddy size similar to those of the colliding particles (Thomas et al., 1999).

3.2.2 Agglomerate strength and hydrodynamic stress

Floc formation and growth process in turbulent flocculation comprises of the lag phase, swift growth phase, and steady state phase or breakup/restructuring phase (He et al., 2015; Bubakova, Pivokonsky, & Filip, 2013; Bemmer, 1979). The overall rate of floc growth at the initial formation phase up to the steady state phase has been suggested to be a balance between collision-induced particle aggregation and floc cohesive strength on one hand, and the rate of floc breakage due to hydrodynamic stress on the other hand (Adachi et al., 2012; Bouyer et al., 2004, 2005; Bridgeman et al., 2009; Coufort et al., 2005; Kobayashi et al., 1999; Soos et al., 2008; Spicer & Pratsinis, 1996).

The aggregate cohesive strength τ is a function of the physicochemical conditions and floc properties while the turbulent hydrodynamic stress σ depends on the design of the aggregation unit (geometry) and the mixing intensity (Adachi et al., 2012; Bouyer et al., 2005; Boyle et al., 2005; Carissimi & Rubio, 2015; Carissimi, Miller, & Rubio, 2007; Carissimi & Rubio, 2005; Kobayashi et al., 1999; Partheniades, 2009). A number of empirical models have been proposed for predicting the maximum hydrodynamic breaking force F_H and the global hydrodynamic stress σ acting on a spherical aggregate in the inertia and viscous domain of turbulence (Table 3.1).

He et al. (2015), in a study of floc strength under turbulent flow conditions derived a mathematical expression for an estimate of the floc binding or cohesive force F_B using fractal dimension approach while Lu et al. (1998) in a similar study presented a theoretical model for the aggregate binding force F_B of spherical monodisperse particles both of which are presented in Eqs. 3.4-3.5. Yuan & Farnood (2010), as well as Jarvis et al. (2005), in their review of aggregate strength and breakage gave an estimate of the floc strength τ and global hydrodynamic stress σ obtained from empirical studies of various types of flocs.

Table 3.1 Empirical and theoretical models of maximum hydrodynamic breaking force

Models	Maximum hydrodynamic force	References
1	$F_{Hmax} = \rho \varepsilon^{\frac{2}{3}} d_f^{\frac{8}{3}}$	(Farinato et al., 1993; Tambo & François, 1991)[a]
2	$F_{Hmax} = \frac{5}{8} \pi \mu d^2 G$	(Soos et al., 2008)[b]
3	$F_{Hmax} = \frac{5}{8} \rho C_I \varepsilon^{\frac{2}{3}} d^{\frac{8}{3}}$	(Bridgeman et al., 2010; Soos et al., 2008)[a]
4	$F_{Hmax} = \left(\dfrac{\pi \rho_f \varepsilon^{\frac{2}{3}} d^{\frac{8}{3}}}{4} \right)$	(He et al., 2015)[a]
5	$F_{Hmax} = \dfrac{\pi \rho \varepsilon}{60 \mu} d^4$	(He et al., 2015)[b]
6	$F_{Hmax} = \dfrac{\rho_{fe} d_f^4}{\mu}$	(Tambo & François, 1991)[b]
7	$F_{Hmax} = \mu G d_f^2$	(Adachi et al., 2012)
8	$F_{Hmax} = \dfrac{\pi \mu d_f^2 G}{2}$	(Rulyov, 2010)
9	$F_{Hmax} = \dfrac{\pi \sigma d_f^2}{4}$	(Bridgeman et al., 2009; Yeung & Pelton, 1996)[b]
10	$F_{Hmax} = \rho^{\frac{1}{3}} \varepsilon^{\frac{2}{3}} d_f^{\frac{8}{3}}$	(Partheniades, 2009)[a]
11	$F_{Hmax} = \dfrac{\rho \varepsilon d_f^4}{\mu}$	(Partheniades, 2009)[b]
12	$F_{Hmax} = d_f d_p \nu G = d_f d_p \sqrt{\varepsilon \nu}$	(Mühle, 1993)[b]

[a] Inertia subrange of turbulence
[b] Viscous subrange of turbulence

$$F_B = 48^{-\frac{2}{3}} \pi^{\frac{5}{3}} d^{\frac{2}{3}} \sigma (\rho_o - \rho_w)^{\frac{2}{3}} d^{\left(1+\frac{D_f}{3}\right)} \qquad (3.4)$$

$$F_B = \frac{(1-P_f) F_{Hmax}}{P_f d_p^2} \qquad (3.5)$$

The global hydrodynamic stress σ due to the shearing action of the fluid motion on the floc as well as the overall mechanical strength of an aggregate τ assuming a uniform floc shape and constant porosity can be expressed mathematically in Eqs. 3.6-3.7 (Coufort, Bouyer & Liné, 2004; Shamlou & Hooker-Titchener, 1993).

In turbulent hydrodynamic flow, the fluctuating motions of the fluid are responsible for shear, tensile, and compressive stress (normal stress) on the flocs, depending on the direction of the fluctuating velocities acting on the aggregates (Liu & Glasgow, 1991, 1997). The exact type and magnitude of the prevailing stress causing aggregate disruption depends mainly on the relation between the floc size and the eddy size or radius, r_w (Mühle, 1993; Wang et al., 2014).

$$\sigma = \mu G = \mu \sqrt{\frac{\varepsilon}{\nu}} \qquad\qquad (3.6)$$

$$\tau = \frac{(1-p)F_B}{pd_p^2} \qquad\qquad (3.7)$$

3.2.3   Fluid-particle interactions and floc stability

The size of aggregates varies from molecular dimensions to a range that is visible to the unaided eye, with the smaller sizes being associated with the primary particles of diameter d_p, while the largest size d_{Fmax} is determined by the balance of floc growth and rupture within the fluid (Bache, 2004; Bridgeman et al., 2009; Partheniades, 1993; Spicer & Pratsinis, 1996). The floc growth and breakage is known to occur simultaneously—growing flocs are subjected to breakage while fragments of broken flocs undergo re-agglomeration until a levelling off of the floc sizes at steady state when the maximum stable size d_{Fmax} is attained (Hogg, 2005; Lick & Lick, 1988; Serra & Casamitjana, 1997; Spicer & Pratsinis, 1996). Figure 3.2 shows that the critical floc size is a balance between the turbulent kinetic energy due to the fluid motion and the intrinsic floc cohesive strength with the rupture of the flocs occurring at a critical or maximum floc size.

There is an increase in floc size as long as the hydrodynamic force due to turbulent shear is less than the floc cohesive or binding force, and after an extended period of time, an equilibrium floc size distribution is reached. In this case either a continued particle or micro flocs attachment to the larger flocs is prevented, or floc breakup

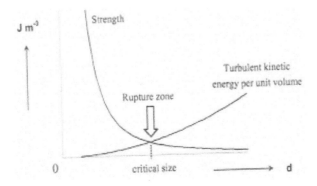

Figure 3.2 Schematic illustration of the equilibrium conditions for the maximum floc size (Reproduced from Bache & Gregory, 2007 with permission © 2007 IWA Publishing).

kinetics balances the turbulence-induced collision (Mühle, 1993; Partheniades, 2009). Increasing the shear rate (agitation or mixing speed) beyond the steady state leads to a change in the mean floc size either due to floc breakage or re-growth (Haralampides, McCorquodale, & Krishnappan, 2003; Lu et al., 1998; McConnachie, 1991; Partheniades, 2009; Spicer & Pratsinis, 1996; Wu, 2008). The steady-state phase is regarded as the balance between floc growth and breakage under a given shear condition. In the viscous subrange, Lu et al. (1998), expressed this phenomenon in terms of the kinetic equation of flocculation as presented in Eq. 3.8. The quantities N_A, N_B, σ_s, τ_s, represent change in the particle number concentration per unit volume for aggregation and breakage, the aggregate shear strength and the shearing stress respectively.

$$\frac{dN}{dt} = -\frac{2}{3}\propto_0 \left(\frac{\varepsilon}{\nu}\right)^{\frac{1}{2}} d_A^3 N_A^2 + \beta_1 \frac{\tau_s}{\sigma_s}\left(\frac{\varepsilon}{\nu}\right)^{\frac{1}{2}} d_B^2 N_B \qquad (3.8)$$

In the analysis of floc breakup under turbulent hydrodynamic conditions, three approaches are normally employed namely: the limiting strength approach, maximum strain rate approach and the maximum floc size approach (Benjamin & Lawler, 2013). The first approach to floc strength analysis which is based on force balances (Mühle, 1993), requires an accurate description of the flow field in order describe the stress acting on the model floc. In this approach for analyzing the breakage phenomenon, under a given agitation condition, the critical condition of floc breakage is attained when the hydrodynamic breaking force F_H is greater than or equal to the floc binding or cohesive force F_B (B ≤ 1).

Consequently, the conceptual form of the floc growth and breakage rate mechanism can be expressed mathematically in Eqs. 3.9-3.10 while Eqs. 3.11-3.12 expressed the critical condition of floc breakage for the inertia and viscous subrange respectively (Adachi et al., 2012; Bouyer et al., 2004, 2005; Bridgeman et al., 2009, 2010; Bridgeman et al., 2008; Coufort et al., 2005; He et al., 2015; Kobayashi et al., 1999; Liu & Glasgow, 1997; Soos et al., 2008). A number of empirical models have been developed for estimating the maximum aggregate size d_{Fmax}. Few of such expressions are presented for the inertia and viscous subrange of turbulence in Table 3.2 (Jarvis et al., 2005; Yuan & Farnood, 2010).

$$\beta_{floc} = \alpha_{ij}\beta_{col_{ij}} - \beta_{br} \qquad (3.9)$$

$$\beta_{floc} = (\alpha_{ijBM}\beta_{BM} + \alpha_{ijSH}\beta_{SH} + \alpha_{ijDS}\beta_{DS}) - \beta_{br} \qquad (3.10)$$

$$B = \frac{\text{Floc binding force}}{\text{Hydrodynamic force}} = \frac{F_B}{F_H} = \frac{48^{-\frac{2}{3}}\pi^{\frac{5}{3}}d^{\frac{2}{3}}\sigma(\rho_o - \rho_w)^{\frac{2}{3}}d^{\left(1+\frac{D_f}{3}\right)}}{\dfrac{\pi\rho_w^2\mu^{\frac{2}{3}}G^{\frac{4}{3}}d^{\frac{8}{3}}}{4}} \qquad (3.11)$$

$$B = \frac{\text{Floc binding force}}{\text{Hydrodynamic force}} = \frac{F_B}{F_H} = \frac{48^{-\frac{2}{3}}\pi^{\frac{5}{3}}d^{\frac{2}{3}}\sigma(\rho_o - \rho_w)^{\frac{2}{3}}d^{\left(1+\frac{D_f}{3}\right)}}{\dfrac{\pi\rho_w d^4 G^2}{60}} \qquad (3.12)$$

Table 3.2 Empirical and theoretical models for estimating maximum floc size under steady state conditions

Models	Maximum floc diameter	References
1	$d_{Fmax} = K' \varepsilon^{-I(I+D_p)}$	(Kramer & Clark, 1999)
2	$d_{Fmax} = 0.68 \rho^{\frac{3}{5}} d_0^{\frac{3}{5}} \varepsilon^{\frac{11}{20}} v^{\frac{9}{20}} F_B^{\frac{3}{5}}$	(Zhu, 2014)[c]
3	$d_{Fmax} = \dfrac{6U(I-p)^{\frac{2}{3}}}{\pi^2 K G \eta d_p^2}$	(Rulyov, 2010)
4	$d_{Fmax} = \dfrac{(F_B / d_p)}{\rho \sqrt{\varepsilon v}}$	(Attia, 1992; Mühle, 1993)[d]
5	$d_{Fmax} = \left[\dfrac{\left(\dfrac{F_B}{d_p}\right) d_p^k}{\rho G v} \right]^{\frac{1}{k+1}}$	(Mühle, 1993)[d]
6	$d_{Fmax} = \left[\dfrac{\left(\dfrac{F_B}{d_p}\right) d_p^k}{\sigma \sqrt{\varepsilon v}} \right]^{\frac{1}{k+1}}$	(Mühle, 1993)[c]
7	$d_{Fmax} = \dfrac{\sqrt{\dfrac{\sigma}{\rho_f}}}{\sqrt{\dfrac{\varepsilon}{v}}}$	(Shamlou & Hooker-Titchener, 1993)[d]
8	$d_{Fmax} = \dfrac{\left(\dfrac{\sigma}{v}\right)^{\frac{1}{3}}}{\dfrac{1}{\varepsilon}}$	(Shamlou & Hooker-Titchener, 1993)[c]
9	$d_{Fmax} = 2\rho^{-\frac{1}{2}} (\varepsilon v)^{-\frac{1}{4}} B_f^{\frac{1}{2}}$	(Lu et al., 1998)[d]
10	$d_{Fmax} = \left[\dfrac{B_f v^{\frac{3}{4}}}{1.9 \rho \varepsilon^{\frac{11}{12}} d_p^2} \right]^{\frac{3}{5}}$	(Lu et al., 1998)[d]

[c] Inertia domain of turbulence
[d] Viscous subrange of turbulence

The fluctuating instantaneous fluid velocities acting parallel to the surface of the floc will induce a local shear stress σ_s on the aggregate in the viscous subrange reaching a maximum value when the floc size is roughly equal to the eddy scale (Kolmogorov micro scale) as illustrated in Figure 3.3 (Attia, 1992).

Similarly, fluctuating fluid velocities normal to the floc surface or dynamic pressure fluctuations acting on opposite sides of an aggregate will results in normal or bulk pressure stress σ_t (tensile or compressive) (Kruster, 1991). In addition, turbulent drag forces F_D acting on the surface of an aggregate which originates from the local motion of the fluid relative to the motion of the aggregates will results in instantaneous surface shear forces and shear stress respectively (Shamlou & Hooker-Titchener, 1993). It has been shown that for flocs in the inertial subrange of turbulence ($d_F > \lambda_0$), tensile stress will predominate causing wholesale fracture or fragmentation (Lu et al., 1998), while for those in the viscous subrange ($d_F < \lambda_0$), shear stress will cause erosion of the particles, floc shell or floc surface as shown schematically in Figure 3.4 (Thomas et al., 1999; Kruster, 1991).

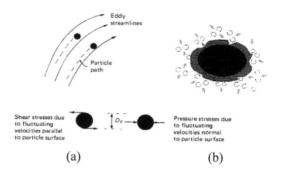

Figure 3.3 Conceptual view of particle-eddy interactions in turbulent field (a) particle size smaller than eddy size (b) eddy size much smaller than particle size (Adapted from Shamlou & Hooker-Titchener, 1993 with permission © 1993 Elsevier).

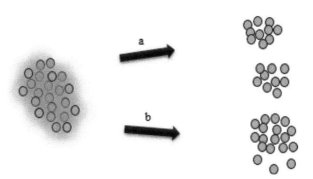

Figure 3.4 Schematic representation of hydrodynamic interactions leading to floc rupture (a) floc splitting or fragmentation (b) surface erosion.

3.2.4 Mechanisms of aggregate disruption

Several mechanisms have been proposed to account for the disruption of aggregates in orthokinetic flocculation (Benjamin & Lawler, 2013; Boyle et al., 2005; Dobias & Von Rybinski, 1999; Gregory, 2006a, 2006b; Jarvis et al., 2005; Mühle, 1993; Neumann & Howes, 2007; Partheniades, 2009; Peng & Williams, 1993; Rulyov, 2010; Thomas et al., 1999; Van Leussen, 2011; Yeung & Pelton, 1996). In addition to floc splitting, fracture or bulk rupture and surface erosion mentioned earlier, several other mechanisms have been proposed to be responsible for floc fragmentation but with little experimental data or theoretical analysis. Such mechanisms include aggregate-aggregate collisions, aggregate-stirrer collisions, as well as collisions with baffles and the mixing tank wall (Bemmer, 1979; Lu et al., 1998).

A number of theoretical and empirical models have been developed to account for the hydrodynamic stress exerted on particle agglomerates and their cohesive strength. Considering a model floc undergoing rupture by fragmentation ($\sigma_t \geq \tau_t$) in the inertia domain of turbulence ($d_F > \lambda_0$), subjected to the shearing action of fluid motion, the turbulent hydrodynamic tensile stress σ_t causing bulk rupture or fragmentation by floc splitting and the aggregate tensile strength τ_t resisting the fluctuating pressure on the floc may be expressed mathematically in Eqs. 3.13-3.14.

$$\sigma_t = 0.49\rho_f(\varepsilon^3 v)^{\frac{1}{4}d_F} \tag{3.13}$$

$$\tau_t = \frac{(1-\rho_f)B_f}{d_p^2} \tag{3.14}$$

Similarly, for a floc rupture due to the erosion of primary particles, micro flocs, or floc shell ($\sigma_s \geq \tau_s$) in the viscous subrange ($d_F < \lambda_0$); the turbulent hydrodynamic shear stress σ_s eroding the surface of the aggregates and the corresponding aggregate shear strength τ_s resisting the viscous shear forces can be expressed mathematically in Eqs. 3.15-3.16 (Lu et al., 1998).

$$\sigma_s = 0.26\rho_f(v\varepsilon)^{\frac{1}{2}} \tag{3.15}$$

$$\tau_s = \frac{(1-\rho_f)B_f}{d_p^2} \tag{3.16}$$

In a related study, Mühle (1993) presented floc breakup analysis on the basis of rheology. The strength of a model floc was described in terms of the surface shear yield strength τ_y resisting floc shell erosion due to pseudo-surface tension force γ_F eroding the particle chain of the outer floc surface in the viscous domain of turbulence ($\sigma_s \geq \tau_y$) (Eq. 3.17). The yield stress approach was also presented by Liu & Glasgow (1997) for calculating the maximum floc tensile yield stress σ_y at which breakage is likely to occur in the inertia subrange ($\sigma_t \geq \sigma_y$) (Eq. 3.18).

$$\tau_y = \frac{F_B}{d_p^2}\left(\frac{\lambda_0}{d_F}\right)^k \tag{3.17}$$

$$\sigma_y = 0.5\alpha\rho\epsilon^{\frac{2}{3}}\left[\frac{2\pi}{d_f}\right]^{-\frac{2}{3}} \tag{3.18}$$

On a similar basis, Attia (1992) presented models for predicting the critical fluid velocity V_c above which there will be floc breakage by estimating floc yield stress σ_y (tensile or compressive) resulting from dynamic pressure acting on the floc (Eqs. 3.19-3.20). The kinetic equations for fragmentation and surface erosion mechanisms are presented in Eqs. 3.21-3.22 based on the empirical results from flocculation in stirred reactor. The quantities N_A, N_B, σ_s, τ_s, represent change in the particle number concentration per unit volume for aggregation and breakage, the aggregate shear strength and the shearing stress respectively (Kramer & Clark, 1999; Lu et al., 1998).

$$V_c = \sqrt{\frac{2\sigma_y}{\rho_f}} \tag{3.19}$$

$$\sigma_y = \frac{1}{2}\rho_f v^2 \tag{3.20}$$

$$-\frac{dN_A}{dt} = KN_A^2 - \beta N_B \tag{3.21}$$

$$\frac{dN}{dt} = \beta_1 \frac{\tau_s}{\sigma_s}\left(\frac{\epsilon}{v}\right)^{\frac{1}{2}} d_F^2 N_F \tag{3.22}$$

Chapter 4

Materials and experimental methods

4.1 TEST MATERIALS

The test materials for the experimental studies consist of the model suspensions (synthetic slurries) and the flow units. All the experiments were carried out using dilute suspension of kaolin and ferric hydroxide precipitates in distilled water. A description of the test materials and their characteristic properties are presented in the following sections.

4.1.1 Kaolin slurry

The kaolin slurry was prepared from high quality coating grade kaolin powder (Caminauer Kaolinwerk GmbH Sachsen, Germany). It contains particle sizes below 36.5 μm with a median size d_{50} of 5.07 μm and density of 2.55 g/cm^3 measured with a pycnometer (Table 4.1). The particle size distribution was determined using Sympatec HELOS H0654 Particle Size Analyzer with the result shown graphically in Figure 4.1. The structure of the kaolin platelet and its pH-dependent surface chemistry is shown in Figure 4.2 which shows the variation in the net surface charge as a function of the pH of the suspension.

4.1.2 Ferric hydroxide slurry

The ferric Hydroxide $Fe(OH)_3$ colloidal precipitates were obtained already pre-mixed from the central storage tank after stirring for 30 minutes. It contains particle sizes below 43.5 μm with a median size d_{50} of 6.14 μm and density of 2.68 g/cm^3 measured with a pycnometer (Table 4.2). The particle size distribution was determined using Sympatec HELOS H0654 Particle Size Analyzer with the results presented graphically in Figure 4.3 while Figure 4.4 shows the structure of the particles.

4.1.3 Synthetic polymers

A number of synthetic polymers Zetag® 7692, Sedipur® CF-2501 (BASF GmbH, Germany), Praestol® 611BC (Stockhausen GmbH, Germany), and Superfloc® C-491, C-492, N-300 (Kemira Oyj, Finland) with different charge densities but similar molecular weights were selected for this study. The selection was made after a preliminary screening based on the knowledge of the substrates and the performance criteria obtained from literature survey [25].

Table 4.1 Physicochemical properties of the kaolin suspension

Parameters	Characteristic values
Streaming potential ΔU, mV	−770
Median particle size d_{50}, μm	5.07
Total solids content, g/l	20
Total dissolved solids, g/l	0.015
pH	6.2
Conductivity, μs/cm	30
Dry solids density, g/cm³	2.55
Specific surface charge, C/g	−0.55
Colour	White

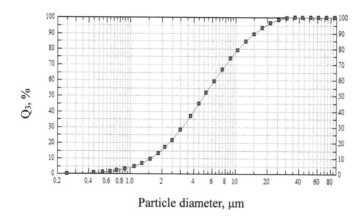

Particle diameter, μm

Figure 4.1 Particle size distribution of the kaolin slurry measured by laser diffraction.

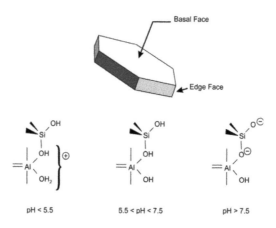

Figure 4.2 The kaolin clay platelet and pH-related surface chemistry of the edge-face. (Reproduced from Hocking, Kimchuk, & Lowen, 1999 © 1999 Taylor & Francis)

Table 4.2 Physicochemical properties of the ferric hydroxide sludge

Parameters	Characteristic values
Streaming potential ΔU, mV	−200
Median particle size d_{50}, μm	6.14
Total solids content, g/l	1.15
Total dissolved solids, g/l	0.0023
pH	7.8
Conductivity, μs/cm	3.23
Dry solids density, g/cm³	2.68
Specific surface charge, C/g	−3.20
Colour	Brown

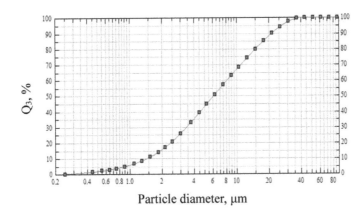

Particle diameter, μm

Figure 4.3 Particle size distribution of the ferric hydroxide slurry measured by laser diffraction.

The charge densities of the synthetic flocculants were determined by colloid titration with standard solutions of Poly-Dadmac–Polydimethyldiallylammonium chloride (0.001 N, 0.04% wt.) and PES-Na–Polyethylene sulphonate (0.001 N, 0.013% wt.) supplied by BTG Instruments GmbH, Germany. The characteristic properties of the flocculants are given in Table 4.3 and their molecular structure in Figure 4.5.

4.1.4   Jar test apparatus

The jar test apparatus for the flocculation experiments consists of an adjustable speed mixer type *Floculateur* 11198 (Fisher Bioblock Scientific, France) with six 1-litre glass jars and a paddle stirrer. The adjustable speed control has a range of 0 to 200 rpm. The mixing jar shown in Figure 4.6 is a cylindrical-shaped glass beaker of 110 mm diameter and 120 mm in height. The paddle stirrer of the jar flocculator is flat and rectangular in shape with its centre located approximately 25 mm above the bottom of the mixing jar. Dilute solutions of the flocculants with different concentrations (0.025-0.1%) prepared from the stock solutions (0.25-0.5%) were injected manually as feed solutions into the pre-mixed suspension using plastic dosing syringe.

Table 4.3 Characteristic properties of the flocculants

Polymer trade name	Supplier	Appearance	Ionic character	Molecular weight $(gmol^{-1})$	Charge activity $(mol\%)$	Charge density (Cg^{-1})
Zetag® 7692	BASF	Granular powder	Cationic	5×10^6	~10	133
Sedipur® CF-2501	BASF	Granular powder	Cationic	6×10^6	~5–10	104
Praestol® 611BC	Stockhausen	Granular powder	Cationic	5×10^6	~3–10	135
Superfloc® C-491	Kemira	Granular powder	Cationic	3.5×10^6	~5	106
Superfloc® C-492	Kemira	Granular powder	Cationic	3.5×10^6	~10	128
Superfloc® N-300	Kemira	Granular powder	Non-ionic	4.5×10^6	~0	−1.997
Poly-Dadmac	BTG	Liquid	Cationic	1×10^5	~100	597.43
PES-Na	BTG	Liquid	Anionic	2.2×10^4	~100	−742.19

Figure 4.4 Microscopic structure of the Ferric hydroxide particles.

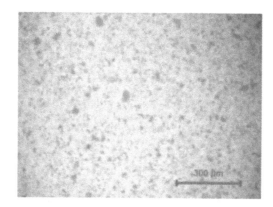

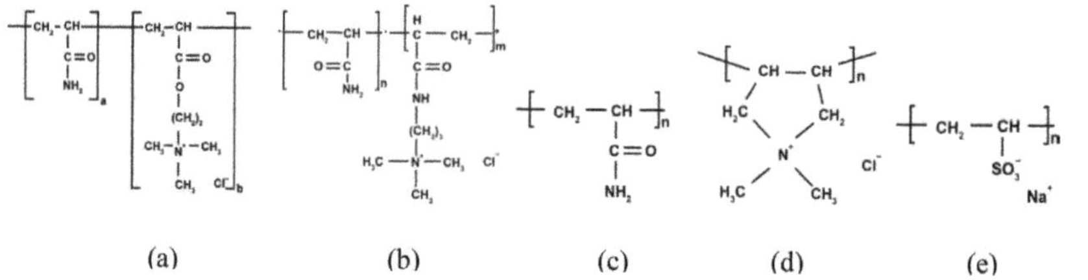

(a)   (b)   (c)   (d)   (e)

Figure 4.5 The molecular structure of the synthetic flocculants (a) Zetag® 7692, Sedipur® CF-2501, Superfloc® C-491—DMAEA-Q or DMAEA-MeCl (b) Praestol® 611BC—MAPTAC (c) Superfloc® N-300—PAM (d) Poly-Dadmac (e) PES-Na.

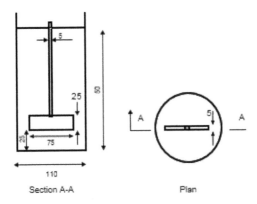

Section A-A   Plan

Figure 4.6 Schematic diagram of the mixing vessel configuration for the jar test (not to scale, all dimensions in millimeters).

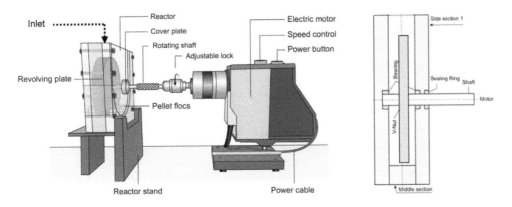

Figure 4.7 Schematic representation of the batch flow vortex reactor.

4.1.5   Batch vortex reactor

The batch flow reactor shown in Figure 4.7 was made from Plexiglass and it consists of a single-cell unit with rotor-stator configuration. The technical data and operating parameters for the reactor are given in Table 4.4. The outer wall consists of a semi-circular cross section that acts as the stator, and axially revolving plates that act as the rotor, which induce flow in an anticlockwise direction. The angular rotation is achieved by means of a motor attached to the outer plate and includes a provision for making speed adjustments. Torque and rotation speed (rpm) measurements were taken directly from the built-in digital display attached to the electric motor. The suspension is loaded to fill one-half of the reactor volume when operated in batch and quasi-continuous mode, such that the pelletization takes place in this area only. The revolving plates are perforated at the center to the outer edge in order to facilitate cleaning of the device and the removal of the supernatant and pellet flocs. There is an opening at the top of the reactor for injecting the flocculant and the suspension.

Table 4.4 Technical data and operating conditions for the batch flow vortex reactor

Overall dimension	180 × 180 × 60 mm
Rotor-stator cavity	3–8 mm
Operating speed	145–165 rpm
Residence time	5–20 minutes
Polymer dosage	3–5 kg/t TS
Feed concentration	1–4%
Separation method	Gravity dewatering ($\phi < 0.5$mm)

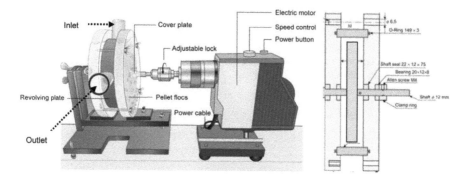

Figure 4.8 Schematic representation of the continuous flow vortex reactor.

Table 4.5 Technical data and operating conditions for the continuous flow vortex reactor

Overall dimension	Cylindrical ($\phi = 136$mm, H = 30 mm)
Rotor-stator cavity	3–8 mm
Operating speed	70–200 rpm
Residence time	5–20 minutes
Polymer dosage	3–5 kg/t TS
Feed concentration	1–4%
Separation method	Gravity dewatering ($\phi < 0.5$mm)

This device was used to simulate the pelleting process in order to determine the optimum process conditions.

4.1.6   Continuous vortex reactor

The continuous flow reactor, also made from Plexiglass, consists of a single-cell unit with rotor-stator configuration as illustrated schematically in Figure 4.8. The technical data and operating parameters for the reactor are given in Table 4.5. The outer wall consists of a circular cross section that acts as the stator, and axially revolving plates that act as the rotor, which induce flow in an anticlockwise direction. The angular rotation is achieved by means of a motor attached to the outer plate and includes a provision for speed adjustments. The filling mode and the operating conditions are similar to the batch vortex reactor with the possibility of utilizing the total volume capacity of the reactor (Table 4.5).

4.1.7 Gravity dewatering chamber

The apparatus made from Plexiglass consists of a cylindrical column equipped with 0.5 mm diameter sieve (Figure 4.9). The flow of water out of the column or the rate of dewatering was controlled by the attached valve at the flow outlet. The dewatering apparatus was used to collect the matured pellets directly from the batch and continuous vortex reactors for subsequent solid-liquid separation by gravity.

4.2 EXPERIMENTAL METHODS

The following sub-sections give a description of the sample preparation and analytical methods employed in this study.

4.2.1 Preparation of substrates and reagents

Colloidal suspension of kaolin (2% wt.) was prepared under laboratory conditions for this investigation by dispersing kaolin powder in deionized water. The mixing was done using a mechanical overhead stirrer Heidolph RZR 2102 (Heidolph Instruments GmbH, Germany) in a 500 mL plastic beaker equipped with baffles. All experiments were conducted using a freshly prepared kaolin suspension. Pre-mixed samples of Ferric Hydroxide $Fe(OH)_3$ precipitates were collected from the central storage tank after mixing for 30-45 minutes. Stock solutions of polyelectrolytes (0.025% wt.) were prepared by dissolving the polymer granules in distilled water and storing them for 24 hrs to allow for the aging of the solution. A working solution was prepared at least one hour before each experiment to obtain a desired concentration. The stock polymer solutions were stored for a maximum of one week.

Figure 4.9 Schematic diagram of the gravity dewatering apparatus.

Standard solutions of acid and base (0.1 M HCl and 0.1 M NaOH) were used for pH adjustments.

4.2.2 Flocculation and sedimentation tests

The flocculation tests were performed according to EN14742:2015 (characterization of sludges—laboratory chemical conditioning procedure). Samples of the synthetic substrates were prepared under laboratory conditions for this investigation. All flocculation experiments were carried out at room temperature (22 ± 0.1°C) using a conventional jar test apparatus Floculateur 11198 (Fisher Bioblock Scientific, France) equipped with time adjustable speed mixer. 250 mL of the working slurry was pre-treated in the flocculator by adding 50 mL each in single addition (Z-7692, CF-2501, 611BC) and 25 mL each in dual additions (Z-7692+N300, CF-2501+N300, 611BC+N300) of dilute solution of the flocculants. A rapid mix of about 45s at 200 rpm was followed by 10 min of slow mixing at 40 rpm to promote the aggregation of flocculated particles and thereafter allowed to settle before further analysis of the sediment and the supernatant.

4.2.3 Surface charge measurements

The surface charge of the synthetic sludge and charge densities of the selected polymers were determined following the SCAN-W 12:04 test method using a colloid titrator 702 SM Titrino (Metrohm AG, Herisau, Switzerland) and a particle charge detector, Mütek PCD 03 (BTG Instruments GmbH, Herrsching, Germany). A 10 mL sample of the slurry (2% wt.) or polymer solution (0.05% wt.) as specified by the manufacturer was placed in the measuring cell of the particle charge detector. The charge in the samples was determined by using either PES-Na (0.001 N, 0.013% wt.) or Poly-DADMAC (0.001 N, 0.013% wt.) as a standard titration solution. The automated system allows the titration parameters such as dosing rates and equivalence end points to be pre-selected for a particular titration regime. The specific charge (μeq/g) is automatically determined by the software. The final charge quantity was obtained thereafter in (C/g) by multiplying the specific charge (μeq/g) by the Faradays constant (96485 C/eq) with an error margin of about ±2%. The determination of the isoelectric point was carried out using PCD 03 by titrating the slurry with a dilute solution of sulphuric acid (0.1 M) until the isoelectric point is reached using standard solution of 0.1 M NaOH for the pH adjustment.

4.2.4 Characterization of the reactor supernatant

A poly-dispersed suspension of kaolin and Ferric hydroxide were pre-treated in the vortex reactors by sequential dual-addition of a cationic high molecular weight polymer with low charge activity and a high molecular weight non-ionic polymer at a mixing speed of 155, 175, and 195 rpm for about 45s was followed by a further mixing time of 5, 10, 15 and 20 mins at slower speed of 145, 165, 185 rpm respectively. Thereafter, samples of the supernatant were withdrawn after 2 minutes of settling for residual turbidity—90° light scattering (Gregory, 2013b) and residual charge measurements using PCD 03 (BTG Instruments GmbH, Herrsching, Germany) and Turbiquant® 3000IR (Merck Millipore GmbH, Hessen, Germany) respectively. The experiments were performed with varying total polymer doses of 3, 4 and 5 kg/t TS.

4.2.5   Flow stream characterization

The flow in both reactors was characterized by means of digital video recording and 2-D Particle Image Velocimetry (PIV) which is an established optical diagnostic technique for fluid flow analysis. The PIV system used in this study is shown schematically in Figure 4.10. It consists of a laser source, LINOS Nano 250-532-100 (Qioptic Photonics GmbH & Co. KG, Göttingen, Germany) providing the illumination to the cross section of the reactor to be studied, two cylindrical lens that spread the laser beam, and a high resolution CMOS camera. The reactor was filled with distilled water kept at 20 ± 1 °C as working fluid and seeded with a mixture (1:1) of silver-coated (S-HGS) and hollow glass spheres HGS (Dantec Dynamics A/S, Skovlunde, Denmark) with an average diameter of 10 μm as the flow tracers for the PIV measurements (Table 4.6). These type of seed particles have been shown to offer good scattering efficiency and a sufficiently small velocity lag (Raffel et al., 2007; S. F. Thomas et al., 2015). Image of the flow containing the seed particles were then captured at 30-100 frames/s

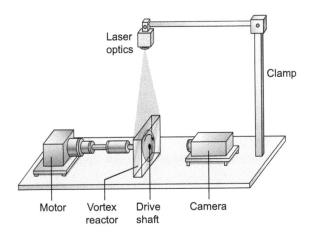

Figure 4.10 Schematic illustration of the Particle Image Velocimetry (PIV) set up.

Table 4.6 Characteristic properties of the seeding particles for the PIV measurements

Seeding particle	Hollow glass spheres	Silver-coated hollow glass spheres
Mean particle size, μm	10	10
Size distribution, μm	2-20	2-20
Shape	Spherical	Spherical
Density, g/cm³	1.1	1.4
Melting point, °C	740	740
Refractive index	1.52	-
Material	Borosilicate glass	Borosilicate glass

using the high resolution camera placed perpendicularly to the laser sheet at different operating speeds and cavity widths.

MATLAB (MathWorks GmbH, Ismaning), MATPIV, GoPro and GIMP software packages were used for the control of the image acquisition and the analysis of the resulting data and to compute the 2-D velocity components u(x,y,t) and v(x,y,t) as well as the vorticity maps as a function of time. Thereafter, the mean flow for the observation plane was obtained by averaging the velocities and vorticities over time. (Kresta & Brodkey, 2004). Cartesian coordinate system was used to describe the flow and the reactor geometry (Utomo, Baker, & Pacek, 2008; Utomo, Baker, & Pacek, 2009).

4.2.6   Agglomerate preparation and characterization

A number of synthetic polymers with high molecular weights and low charge densities: Superfloc® C-492, C-491 and N-300 (Kemira Oyj, Finland) which were selected based on preliminary screening of potential flocculants and physicochemical optimisation are used as bridging materials. The working suspensions which consists of a poly-dispersed kaolin and Ferric hydroxide suspensions were pre-treated in the shear device by adding 8 ml. each of dilute solution of polyelectrolyte with different concentrations as feed solutions (0.2, 0.3, 0.4, 0.5 and 0.6 g/l), after which the suspension was continuously stirred at a uniform shear rate G for a sufficient time to allow for the formation of stable pellets.

Subsequently, the flocs were separated from the supernatant by gravity drainage through a 0.5 mm diameter sieve. The dry matter content of the pellet flocs was determined according to EN15934:2012 (sludge, treated biowaste, soil and waste— calculation of dry matter fraction after determination of dry residue or water content) by drying in a thermostatically-controlled oven with forced air ventilation at 105°C for 24 hrs to a constant mass. Thereafter, the dry solids content was calculated according to Eq. 4.1. The particle removal efficiency R(x), which is a measure of the efficiency of particle removal from the slurry onto the pellet flocs, was determined according to Eq. 4.2.

$$DS, \% = \frac{\text{Dry mass of pellet after 24 hrs of drying}}{\text{Mass of green pellet flocs}} \times 100\% \qquad (4.1)$$

$$R(x), \% = \frac{\text{Dry mass of pellet flocs filtered through 0.5 mm sieve}}{\text{Dry mass of kaolin particles in suspension}} \times 100\% \qquad (4.2)$$

The compressive strength of individual dewatered pellets was determined by single particle uniaxial compression or crushing test shown schematically in Figure 4.11. The test apparatus is a ProLine universal test machine (Zwick/Roell GmbH, Germany) which consists of two parallel plates and a vertical load. In this test, a single dried pellet is compressed between two plates and the load or force causing the fracture is recorded. Force measurements are thereafter correlated with the agglomerate strength and nominal axial strain rate ϵ by using the expression in Eq. 4.3 and Eq. 4.4 where τ_c is the agglomerate compressive strength, F_{max} the

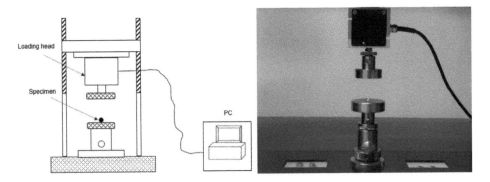

Figure 4.11 Schematic diagram of the single pellet uniaxial compression test set up.

maximum applied load at the point of fracture, $\in(t)$ the strain rate, v the test velocity, and d_p the diameter of the agglomerate (Adams & McKeown, 1996; Barbosa-Canovas et al., 2005).

$$\tau_c = \frac{4F_{max}}{\pi d_p^2} = \frac{1.2727F_{max}}{d_p^2} \tag{4.3}$$

$$\in(t)\frac{\delta\epsilon}{\delta t} = \frac{\delta d_p}{d_p\delta t} = \frac{v(t)}{d_p} \tag{4.4}$$

Chapter 5

Design concept and process description

In the technical design of the processing technique and the test apparatus, the main aim is to generate the desired flow fields in order to realize the rolling and collisional effects necessary for the formation of pellet flocs. This is achieved by selecting an appropriate stirrer-vessel system and identification of the optimum process conditions. The following important parameters were considered in the preliminary process design and the choice of simulation units.

- The geometry of the mixing vessel (shape, size, dimensioning etc.)
- The configuration of the stirrer (shape, diameter, thickness etc.)
- The residence time T
- The velocity gradient or shear regime G
- The type of polymer and optimum dose

On the basis of these identified factors, the structural design of the flow units, geometries and operating conditions are given in sections 4.1.5 and 4.1.6. The processing technique relies of the propagated flow vortex and the reactor geometry to effect the pelletization. A rotating plate of variable thickness (14–24 mm) placed at a distance of 1 mm from the reactor wall on either side acts as both the mixer and flow inducer as illustrated in Figure 5.1.

5.1 STRUCTURE FORMATION AND HYDRODYNAMIC CONDITIONS

5.1.1 Theoretical description of the agglomeration process

The design principle of the pre-treatment technique as a technical proof of concept is based on shear-assisted phase separation using turbulent vortex flow within a rotor-stator cavity. In practice, the rotor-stator configuration of the system allows an infinite number of cells arranged horizontally to be in operation simultaneously thereby saving cost and considerably increasing the throughput. The flow within the reactor cell can be characterized both quantitatively and qualitatively by forced vortex rotational flow where the fluid particles rotate about a fixed axis like a solid body with a constant angular velocity ω and a vorticity 2ω for a cylindrical coordinate

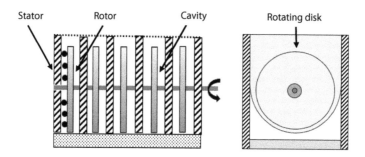

Figure 5.1 Model of the batch vortex reactor showing the stirrer-vessel configuration.

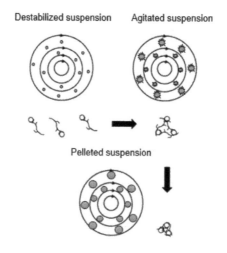

Figure 5.2 Conceptual model of structure formation process and flow pattern in plane circular vortex flow field.

system (Eq. 5.1) (Biswas & Som, 2004; Singh, 2012; Ward-Smith & Massey, 2006). The fluid mass is made to rotate by the application of external power source via the rotating shaft. The flow streamline is characterized by plane circular or cylindrical vortex in the case of full-volume continuous reactor and spiral vortex motion in the case of the batch reactor.

In the plane circular or cylindrical forced vortex motion shown in Figure 5.2, the flow streamlines resembles concentric circles where the tangential velocity V_θ is directly proportional to the radius of curvature while the radial velocity is equal to zero (Eqs. 5.2-5.3). The spiral forced vortex motion on the hand is a superimposition of the purely radial flow (inward or outward) on a plane circular vortex motion as illustrated in Figure 5.3 (Biswas & Som, 2004). In either of the two types of vortex flow, mechanical energy supplied by the application of an external torque is required to maintain the vortex flow. The observed turbulent flow in both reactors results in the formation of a plane circular and spiral vortex flow fields on

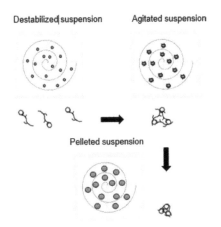

Destabilized suspension   Agitated suspension

Pelleted suspension

Figure 5.3 Conceptual model of structure formation process and flow pattern in spiral vortex flow field.

either side of the reactor, providing the necessary rotational effect and mass transfer (Visscher et al., 2013). The vorticity ξ which can be defined in both 2-D and 3-D, is an indication of the spin, curl or rotation of the fluid as it moves, while the tip velocity V_{tip} of the flow inducer which is a measure of the tangential velocity V_θ gives an indication of the strength of the flow vortex or swirl generated in the reactor (Bridgeman et al., 2009; Marshall & Bakker, 2004; Thomas et al., 2015). Under the prevailing operating conditions, the flow in the reactor is fully turbulent ($R_e > 10^4$) where R_e is the Reynolds number which indicates the degree of instability in the flow (Zlokarnik, 2008).

$$\xi = (\nabla \times \vec{u}) = \frac{\partial V_\theta}{\partial r} - \frac{\partial V_r}{\partial_\theta} + \frac{V_\theta}{r} = \omega - 0 + \omega = 2\omega \tag{5.1}$$

$$V_\theta = \omega r = 2\pi N r = \pi N D \tag{5.2}$$

$$V_r = 0 \tag{5.3}$$

5.1.2   Hydrodynamics and flow pattern in the flow units

The stream pattern of the turbulent flow and its effect on the pre-treatment process was observed both visually and with the aid of a digital camera. A series of short video recordings for a duration of approximately 60s were made during the pelleting process which were thereafter analysed. The path of the agglomerates in the captured video was used to create the flow stream pattern of the observation field. This is based on the assumption that the path of the agglomerates aligned closely with the fluid motion. The turbulent flow in the batch reactor results in the formation of a circular vortex flow field (elliptical shape) on either side of the reactor, providing both the necessary rotational and collisional effects as well as enhanced mass transfer and energy transport that are crucial for the formation of pellet flocs as illustrated in Figure 5.4 (Oyegbile, Ay, & Narra, 2016; Spicer, 1997; Visscher et al., 2013).

In the case of the continuous vortex reactor, an axis-symmetrical spiral vortex flow field was observed on either side of the reactor providing the necessary flow conditions that are crucial for the formation of pellet flocs as shown in Figure 5.5 (Spicer, 1997). The speed of rotation required to keep the formed pellet flocs in motion was observed to be significantly higher in the batch reactor. By contrast, the pellets motion within the flow field can be sufficiently maintained at a rotation speed as low as 70 rpm in the continuous reactor. This is an important advantage in terms of the power consumption and pellet stability as the flocs are susceptible to breakage at higher agitation speed.

5.1.3   Experimental analysis of the hydrodynamics and vortex pattern

The hydrodynamics and the vortex pattern in both reactors were investigated experimentally using PIV technique. An understanding of the nature of the flow stream within the reactors' cavity will be extremely useful in the optimization of the process engineering conditions. The time averaged flow velocity vectors and vorticities over the entire duration of the captured video (60 s) were computed for the observation field. Measurements were taken in the batch reactor at two different water levels (at the mid-point and below the shaft) while full-volume was used in the case of continuous reactor. Flow field were computed at three points in the continuous reactor along the rotor-stator cavity (near the

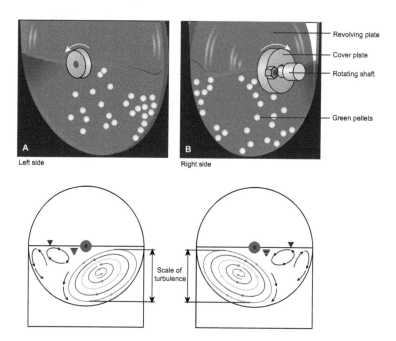

Figure 5.4 Schematic illustration of the relative motion of aggregates and fluid in the batch vortex flow from video recordings.

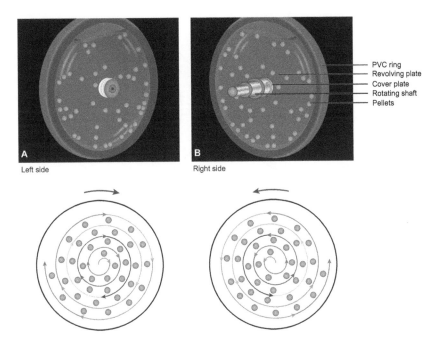

PVC ring
Revolving plate
Cover plate
Rotating shaft
Pellets

Left side

Right side

Figure 5.5 Schematic illustration of the relative motion of aggregates and fluid in the continuous vortex flow from video recordings.

rotor, at the centre, and at the stator) and at the centre in the batch reactor. The results of the time averaged velocity profile and vorticities for both reactors and their effect on the pelleting process are presented and discussed in the following sections.

5.1.3.1 *Description of the experimental setup and limitations*

In view of the nature and complexity of the flow devices, a number of limitations were encountered in the course of the experimental hydrodynamic measurements as shown schematically in Figure 5.6. The first limitation involves the appearance of dark spots on either sides of the measurement plane due to the refraction of the laser beam as it passes through the reactor wall into the simulating fluid (Figure 5.6a). A solution was devised in the subsequent data processing for the PIV computations by masking out this region. In the second limitation, there is a high concentration of the laser beam in the lower part of the measurement plane (at the lower part of the cavity) especially in the continuous reactor and as such, this region was blackened in order to minimize this effect (Figure 5.6b).

The third limitation arises from the appearance of air bubbles and shadow above and below the rotating shaft during the measurements. The air bubbles observed in the continuous reactor was as a result of the pressure differentials (inside and outside of the reactor) while the shadow was due to the laser light passing through the small injection opening at the top of the reactor (Figure 5.6c). Consequently, this region had to be masked out in the subsequent PIV computations. The last limitation involves

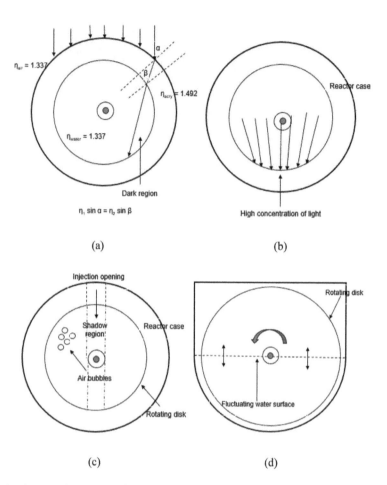

$n_i \sin \alpha = n_2 \sin \beta$

(a)

(b)

(c)

(d)

Figure 5.6 A schematic description of experimental limitations encountered in the flow measurements (a) dark region (b) bright spots (c) air bubbles and shadow (d) laser sheet fluctuations.

the rapid fluctuations of the water surface and laser sheet in the batch reactor which increases with the agitation speed. Unfortunately, no solution was found to this problem and the flow measurements could not be performed at a rotation speed higher than 100 rpm (Figure 5.6d). Cartesian coordinate system (with its centre point at the middle of the rotating shaft) was used to describe the flow and the reactor geometry.

5.1.3.2 Effect of agitation speed on the flow stream and vortex pattern

The effect of the agitation speed on the flow stream and vortex pattern was determined with measurements taken at different agitation speeds in both reactors. However, due to rapid fluctuations in the laser signal as a result of flow instability at the water surface, the flow measurements in the batch reactor could not be performed at a rotation speed higher than 100 rpm. This problem was not encountered in the continuous reactor as the flow was completely enclosed within the reactor. The results of the time averaged velocity vectors and vorticity map at different agitation speed in the batch reactor are presented in Figure 5.7 and Figure 5.8 where the positive and negative

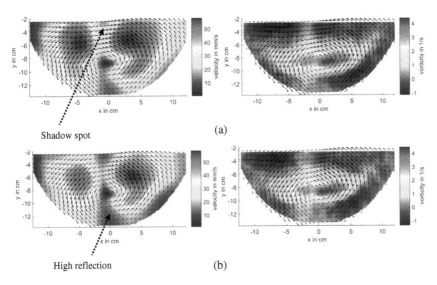

Figure 5.7 2-D flow stream and vortex pattern showing the averaged velocity vector field, mm s^{-1} and vorticity map, s^{-1} with water level below the shaft of the batch reactor (a) at 80 rpm (b) at 100 rpm.

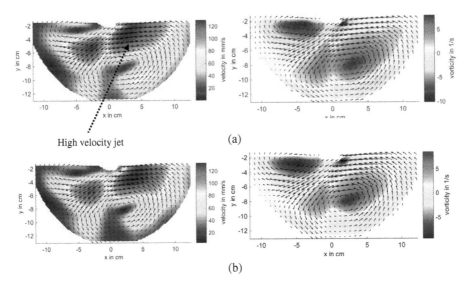

Figure 5.8 2-D flow stream and vortex pattern showing the averaged velocity vector field, mm s^{-1} and vorticity map, s^{-1} with water level at the middle of the batch reactor (a) at 80 rpm (b) at 100 rpm.

values indicate rotation in anticlockwise and clockwise directions respectively. The averaged velocity and vorticity pattern shows two clear regions characterized by high velocity and vorticity values.

The flow at lower water level (below the shaft) results in the formation of a core vortex concentrated at the centre of the reactor. In contrast to the lower water level, at a higher water level (at the middle of the tank), the centre of the vortex shifts to the right side of the reactor while a smaller vortex develops at the left of the reactor due to a high velocity recirculating jet. This observation in the vortex pattern seem to correlate qualitatively with the visual observation shown in Figure 5.4 which lends credence to the visual observation. In terms of the averaged velocity, the magnitude of the velocity appears to increase away from the centre of the vortex as expected, while that of the vorticity has a maximum value at the core of the vortex which is consistent with the theoretical model. This implies that the rotational effect is at the maximum around this region while the flow velocity increase outwardly.

The results of the time averaged velocity and vorticity map at different agitation speed in the continuous reactor at the centre of the cavity are presented in Figure 5.9 and Figure 5.10. The flow fields in these figures show that the velocity magnitude increases with the agitation speed and away outwardly from the rotating shaft with the maximum values at the edge of the field. In terms of the vorticity as expected, there is an even distribution of the vorticity at all operating speed within the field except around the edges where there are very low vorticity values (because of the dark spots and bubbles) and this area of low vorticity increases slightly with the operating speed. However, the values obtained around this region are unreliable due to the experimental limitations earlier mentioned.

5.1.3.3   Effect of reactor geometry on the flow stream and vortex pattern

In order to determine the influence of the geometries on the flow stream and vortex pattern, a comparison of the flow field and vortex patterns for both reactors at similar operating speeds was carried out. In terms of the magnitude and distribution of the velocity and vorticity, these parameters appear to be higher and more evenly distributed in the continuous reactor. This partly explains why the continuous reactor can be operated even at a much lower operating speed (70 rpm). In the batch reactor, high velocity magnitude and vorticity appears to be concentrated at the middle of the vortex with a strong high velocity recirculating jet creating a diverging flow as the jet hits the boundary (Figure 5.8) while in the continuous reactor the flow is more evenly distributed. The core vortex in the batch reactor is on a smaller scale and therefore much stronger than that of the continuous reactor which covers a bigger scale in the observation plane.

5.1.3.4   Description of the flow stream and vortex pattern along the rotor-stator cavity

In order to determine the variations in the flow stream and vortex pattern along the rotor-stator cavity, the flow in the continuous reactor was characterized experimentally with measurement taken at three positions along the cavity (near the rotor, at the center and near the stator). Figure 5.11 and Figure 5.12 show the averaged velocity

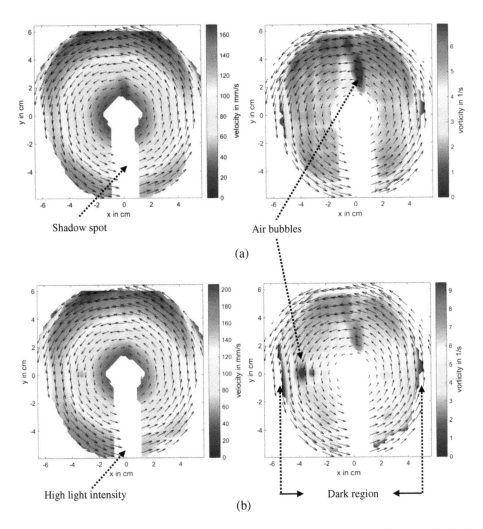

Figure 5.9 2-D flow stream and vortex pattern in continuous reactor showing the averaged velocity vector field, mm s^{-1} and vorticity map, s^{-1} at the center of cavity. (a) at 70 rpm (b) at 90 rpm.

and vorticity fields at the lowest and highest operating speeds for three points along the cavity. The averaged velocity and vorticity pattern of the measurement domain shows an axis symmetrical flow stream characterized by the high velocity and vorticity values. The 2-D vortex pattern appears to form concentric circles around the rotating shaft. There are a few visible dark spots at the middle and edge of the measurement plane due to the experimental limitations.

The magnitude of the averaged velocities increases outwardly from the centre of the concentric circles irrespective of the axial position. However, the highest tangential velocities appear to be at the centre of the cavity and near the rotor with lowest values at the stator. In addition, the vortex appears to form an outward spiral at the

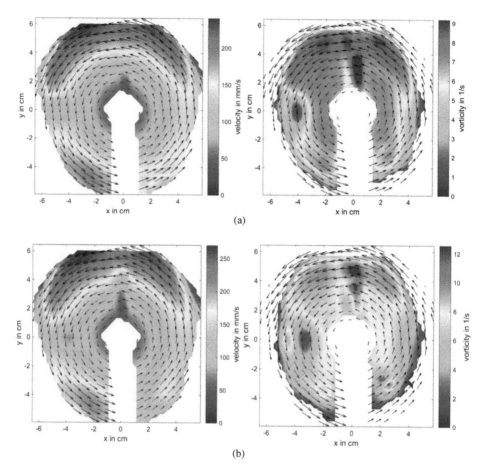

Figure 5.10 2-D flow stream and vortex pattern in continuous reactor showing the averaged velocity vector field, mm s^{-1} and vorticity map, s^{-1} at the center of cavity. (a) at 110 rpm (b) at 130 rpm.

rotor and an inward spiral at the stator but this effect can be better captured in 3-D PIV measurements. There appears to be high signal fluctuations at a rotation speed higher than 140 rpm which prevents the fluid velocity to be accurately measured beyond this point. In terms of the fluid vorticity, the measurement plane is characterized by regions of low vorticity (near the reactor wall) and high vorticity (towards the shaft) although the values are less accurate due to the experimental limitations. The highest vorticity distribution in the flow field is observed near the rotor.

5.1.3.5 Process relevance of the flow conditions

A description of the flow filed and vortex pattern provides a better understanding of the hydrodynamics in terms of the magnitude and distribution of the averaged velocity vector and vorticity in the observation field (pelleting area). It also serves as

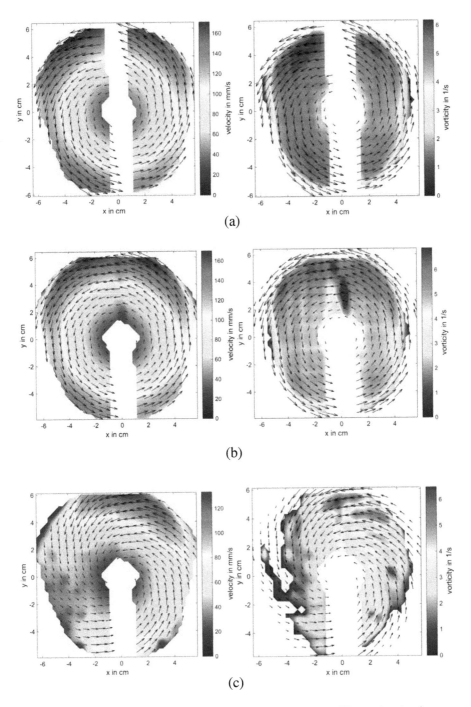

Figure 5.11 2-D flow stream and vortex pattern in continuous reactor at 70 rpm showing the averaged velocity vector field, mm s^{-1} and vorticity map, s^{-1} (a) near the rotor (b) at the center of cavity (c) near the stator.

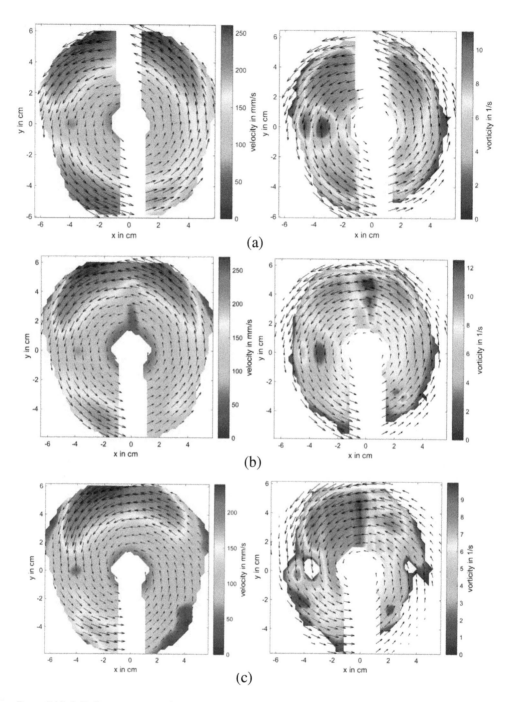

Figure 5.12 2-D flow stream and vortex pattern in continuous reactor at 130 rpm showing the averaged velocity vector field, mm s^{-1} and vorticity map, s^{-1} (a) near the rotor (b) at the center of cavity (c) near the stator.

a basis for the qualitative validation of the visual observation and in the quantitative description of the flow stream. However, the exact flow stream and vortex pattern at the optimum agitation speed for the pelleting process in both reactors (145–195 rpm) could not be measured due to high fluctuations in the laser signal. The high turbulence condition in both reactors makes such measurements quite challenging. Nevertheless, the PIV measurements at a maximum speed of 100 and 130 rpm for the batch and continuous reactor respectively provides an insight into the flow conditions and demonstrates a very good qualitative agreement with the visual observation. It provides a broad picture of the variations in the flow parameters in both reactors with respect to the agitation speed, geometry, and cavity gap.

Chapter 6

Mixing and hydrodynamic analysis

6.1 THEORETICAL ANALYSIS OF THE FLOW CONDITIONS

The flow conditions and the energy input has been shown to influence the mixing of the slurry particles with the flocculant molecules as well as the pre-treatment process as discussed in sections 1.1–3.3. The Hydrodynamic conditions in the reactors were characterized using empirical and theoretical models available in scientific literature (Boller & Blaser, 1998; Bouyer, Escudié, & Liné, 2005; Bridgeman et al., 2009; Edzwald et al., 1997; Lu et al., 1998; Mhaisalkar, Paramasivam, & Bhole, 1986; Thomas et al., 2015). The parameters of interest are those that describe the nature of the flow, hydrodynamic condition and floc stability in the reactor such the power consumption (P), Reynolds number (R_e), kinetic energy transfer or dissipation rate (ε), mixing regime and mode of aggregate failure (λ_0), shear rate (G), strength of the vortex (V_{tip}) and mixer capacity and efficiency (N_p). These parameters are computed and presented in Table 6.1 and Table 6.2.

6.2 ENERGY INPUT AND POWER CONSUMPTION

In order to determine the mass transport and transfer rates in the sheared fluid, the kinetic energy dissipation rate ε in the system must be determined from the energy input into the system (Logan, 2012). The mixing and the reaction of the fluid constituents (flocculant molecules and slurry particles) is achieved by external power input into the system. The energy carrying eddies of dimension $\Lambda \sim d_A$ introduced into the system by the flow agitator will be consequently dissipated into heat by the smallest eddies of length scale λ_0 due to the viscous shear. The power input into the system was calculated from the digital torque readings displayed on the electric motor according to the equation in Table 6.1 and Table 6.2. The kinetic energy dissipation rate ε appears to increase with the agitation speed in both reactors with the continuous reactor showing a higher energy dissipation rate and consequently higher mass transfer rate as shown in Figure 6.1.

6.3 FLUID FLOW AND MIXING CHARACTERISTICS

A description of the mixing characteristics in terms of the turbulence and hydrodynamic indicators is based on the theoretical models available in literature.

Table 6.1 Hydrodynamic characteristic values as a function of the rotational speed for the batch reactor

Hydrodynamic parameters		Mixing speed (rpm)		
Quantities	Empirical models	145	165	185
T_q (Nm)		0.229	0.236	0.244
P (W)	$P = T_q \omega = \dfrac{2\pi n T_q}{60}$	3.48	4.08	4.73
R_e (-)	$R_e = \dfrac{\rho N D^2}{\mu}$	48792	55522	62252
N_p (-)	$N_p = \dfrac{T_q \omega}{\rho N^3 D^5}$	5.30	4.22	3.47
G (s^{-1})	$\bar{G} = \sqrt{\dfrac{P}{\mu V}}$	6289	6811	7333
ε (m^{2}s^{-3})	$\varepsilon = \dfrac{P}{\rho V}$	36.24	42.49	49.26
λ_0 (μm)	$\lambda_0 = \left[\dfrac{v^3}{\varepsilon}\right]^{\frac{1}{4}}$	12.07	11.60	10.18
V_{tip} (ms^{-1})	$V_{tip} = \omega r = \pi N D$	1.0330	1.1754	1.3179
ξ (s^{-1})	$\xi = 2\omega = 4\pi N$	30.3814	34.5719	38.7624
σ (Nm^{-2})	$\sigma = \mu G$	5.7615	6.2393	6.7177

The operating conditions roughly correspond to the shear rate in the range of 6289 to 7681 s^{-1}. The flow is induced by the tangential velocity V_{tip} with a range of 1.03 to 1.39 ms^{-1} at the respective operating conditions. However, by contrast, a lower time averaged flow velocity and vorticity values were recorded in the experimental analysis when compared to the theoretical estimate of the tangential velocity and vorticities at the same operating speed. This might be attributed the boundary effects causing friction across the rotor-stator cavity resulting in a slower fluid flow. The optimum height of the fluid in the batch process is approximately at half of the reactor volume while in the continuous reactor, the height of the fluid may be adjusted according to the process conditions. The flow in both reactors characterized by high Reynolds number R_e is fully turbulent ($R_e > 10^4$) for all the operating conditions with a range of 48792 to 65617 and this increases with the agitation speed as shown in Figure 6.2.

The power number—Reynolds number relationship (*power curve*) is an important criteria in determining the suitable stirrer capacity for a particular mixing regime (Godfrey & Amirtharajah, 1991). At a high Reynolds number (i.e. $R_e > 10^4$) the flow is fully turbulent and the mixing is rapid due to the motion of the turbulent eddies (Edwards, Baker, & Godfrey, 1997). The power curve for the reactors shown in Figure 6.3 indicates that the mixing is fully turbulent in both reactors, efficient and is theoretically assumed

Table 6.2 Hydrodynamic characteristic values as a function of the rotational speed for the continuous reactor

Hydrodynamic parameters		Mixing speed (rpm)		
Quantities	Empirical models	155	175	195
T_q (Nm)		0.237	0.247	0.254
P(W)	$P=T_q\omega=\dfrac{2\pi n T_q}{60}$	3.85	4.53	5.19
R_e (-)	$R_e=\dfrac{\rho N D^2}{\mu}$	52157	58887	65617
N_p (-)	$N_p=\dfrac{T_q\omega}{\rho N^3 D^5}$	4.80	3.92	3.25
G (s⁻¹)	$\bar{G}=\sqrt{\dfrac{P}{\mu V}}$	6615	7176	7681
ε (m²s⁻³)	$\varepsilon=\dfrac{P}{\rho V}$	40.09	47.17	54.05
λ_0 (μm)	$\lambda_0=\left[\dfrac{v^3}{\varepsilon}\right]^{\frac{1}{4}}$	11.77	11.30	10.92
V_{tip} (ms⁻¹)	$V_{tip}=\omega r=\pi ND$	1.1042	1.2467	1.3892
ξ (s⁻¹)	$\xi=2\omega=4\pi N$	32.4766	36.6672	40.8577
σ (Nm⁻²)	$\sigma=\mu G$	6.0601	6.5736	7.0367

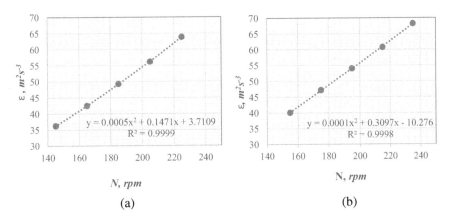

Figure 6.1 Mean kinetic energy dissipation rate ε as a function of the agitation speed N (a) batch reactor (b) continuous reactor.

to be homogenous throughout the vessel. At this scale of mixing (micro mixing), the eddy scale similar to the size of the particles are responsible for the particle-particle inter-actions. In terms of the mixing characteristics, the power curve of both reactor appears to correlate well with that of typical mixers found in literature which indicates that the

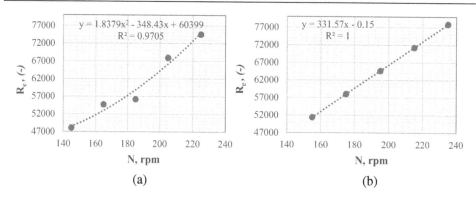

Figure 6.2 Reynolds number R_e as a function of the agitation speed N (a) batch reactor (b) continuous reactor.

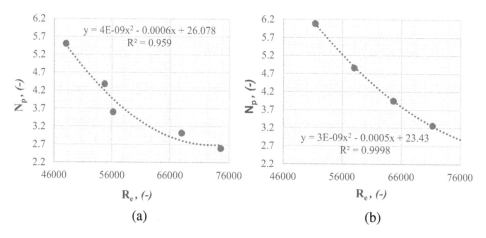

Figure 6.3 Relationship between power number N_p and Reynolds number R_e (a) batch reactor (b) continuous reactor.

mixing conditions is efficient in the reactors. The minor difference in the shape of the curve might be attributed to different frictional losses in the reactors (i.e. turbines, paddles, and propellers) than that of the revolving plate in the reactors (Edwards et al., 1997; Godfrey & Amirtharajah, 1991; Thoenes, 1998).

6.4   MICRO AND MACRO MIXING

The Kolmogorov theory of locally isotropic turbulence (Kolmogorov, 1991a, 1991b) serves as the basis for theoretical interpretation of the macro and micro mixing in the vortex reactors. In practice however, the turbulence in most mixing vessels is anisotropic in nature with spatial and temporal variations (Bridgeman et al., 2008; John Bridgeman et al., 2010). The aggregation and destruction of the agglomerates is strongly influenced by the intensity and the nature of the turbulence. The dimension of the largest eddies responsible for macro mixing is determined by the dimension of

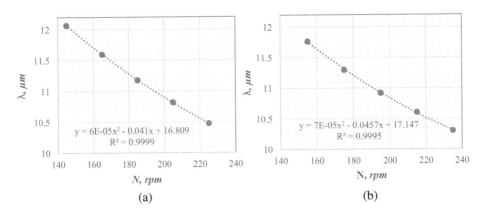

Figure 6.4 Kolmogorov microscale λ_0 as a function of the agitation speed N in batch reactor (a) batch reactor (b) continuous reactor.

the flow inducer d_A, in this case the diameter of the rotating plate. The macro scale of turbulence Λ characterizes the macro structure of the turbulent eddies in the reactor. The micro scale of turbulence on the other hand describes the size of the turbulent eddies at the micro scale of mixing and is an indicator of micro mixing as well as the mode of agglomerate failure. The micro scale of turbulence λ_0 decreases as the agitation speed increases in both reactors as shown in Figure 6.4.

Efficient micro mixing is obtained in both reactors with approximately 70 % of the particles smaller than the Kolmogorov length scale λ_0 at the respective operating speed. In addition, the distribution of the turbulence can be assumed to be more homogenous than in stirred tank reactor due to the small gap between the agitator tip and the reactor wall. However, this was not experimentally verified in this study.

In contrast to the vortex reactors, the turbulence intensity in the conventional tank stirred tank reactor has been shown to vary throughout the mixing vessel with a maximum value near the stirrer head (André, Oshinowo, & Marshall, 2000; Bakker & Oshinowo, 2004; Essemiani & De Traversay, 2002; Korpijärvi et al., 2000).

Chapter 7

Optimization of the micro processes

7.1 OPTIMIZATION OF THE PHYSICOCHEMICAL PROCESS

The first step in the physicochemical optimization process is to define specific performance criteria as well as indicators—characteristic parameters of flocculation efficiency. These criteria as well as the chosen indicators must be tailored to the suspension and the existing or proposed method of treatment in order to obtain the desired floc's characteristics at a reasonable cost (Bratby, 2006; Moody & Norman, 2005). The "optimum" flocculation condition determined either by direct in-situ observation such as particle counting or by other indirect indicators (turbidity, optical density, sediment volume, etc.) will roughly correspond to the minimum characteristic values of the observed parameters as illustrated in Figure 7.1 (Gregory, 2013b). The optimum point is a balance of factors reducing and increasing the observed values as the polymer dose increases.

The physicochemical conditions influencing the process of aggregate formation was studied by conducting a number of jar test experiments and surface charge measurements. This analysis is a prelude to the actual pre-treatment process and provides information on the slurry polymer-dose response which served as a guide in the selection of appropriate conditioning chemicals and dosages and in subsequent performance optimization of the pre-treatment devices. The physicochemical parameters of interest include the polymer dose and type, mixing regime, addition sequence and shear regime. In view of a large array of polymers available, a preliminary screening based on published literature was carried out to narrow the choice of flocculants for the physicochemical test (Sievers et al., 2008; Walaszek & Ay, 2005). The assessment of the flocculation efficiency was carried out on the basis of surrogate indicators such as sludge volume index I_{SV} or specific volume of sediment and supernatant turbidity; which were chosen based on the knowledge of the slurry characteristics and the pre-treatment technique (pelleting flocculation) as well as subsequent solid-liquid separation method employed in this study (gravity dewatering). The chosen flocculation performance criteria (clear effluent, maximum sediment compaction or minimum sludge volume) as well as their performance indicators (turbidity, sludge index or specific volume of sediment) must reflect a trade-off between best possible separation efficiency and lowest possible cost of conditioning chemicals.

In the physicochemical tests, slight deviations were allowed from the expected process conditions in the pelleting process. For instance, the applied shear was

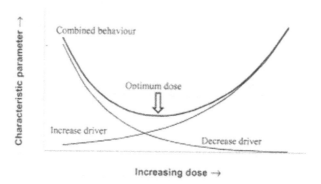

Figure 7.1 Schematic illustration of the parametric behavior-dose plot showing optimum flocculation condition (Reprinted from Bache & Gregory, 2007 with permission © IWA Publishing).

slightly higher than in actual pelletization tests at 200 rpm in the jar testing device as compared to a maximum shear rate of 195 rpm in the flow units. Conversely, the duration of the shear was longer in the pelletization test at approximately 20 minutes as compared to 10 minutes in the physicochemical test. This was necessary in order to allow sufficient time for the densification of the formed pellets. Similarly, the compaction of the agglomerated flocs in the jar flocculator was primarily due to gravitational forces while in the flow units, externally-induced mechanical forces are primarily responsible for the densification of the classical random flocs to form pellet agglomerates. The results of the physicochemical investigation are presented in Table 7.1.

7.1.1   Single and dual polymer conditioning

In the single polymer conditioning, 250 mL of the working slurry was pre-treated in the flocculator with 50 mL of dilute solution of polyelectrolytes (Z-7692, CF-2501, 611BC) of varying concentrations (0.1–0.8 g/L) in order to obtain a dosage range of 1–8 kg/t TS. A rapid mix of about 45s at 200 rpm was followed by 10 min of slow mixing at 40 rpm to promote the aggregation of flocculated particles. The suspension was then transferred into a 250 mL graduated cylinder and allowed to settle undisturbed.

Samples of the supernatant were drawn with a pipette from 2 cm below the surface for residual turbidity measurements using Turbiquant® 3000IR nephelometer (Merck Millipore GmbH, Hessen, Germany), while the sludge volume index or specific volume of sediment I_{sv} was determined according to Eq. 5.4 after 30 mins and 24 hrs of quiescent settling respectively where V_s is the volume of sediment after 30 min and 24 hrs of settling, and is ρ_T the suspended solids concentration of the slurry.

$$I_{SV} = \frac{V_s}{\rho_T}$$

(5.4)

Table 7.1 Summary of results for flocculation tests in single and dual-polymer treatment

Polymer type	Performance indicators	Polymer dose (kg/t TS)				
		1	2	4	6	8
Z-7692	Sludge volume index $I_{SV\,30}$	6.6	5	5	6.4	6.6
	Sludge volume index $I_{SV\,24}$	6.2	4.8	4.8	6.4	6.2
	Turbidity	6.65	12.92	38.13	63.32	64.39
CF-2501	Sludge volume index $I_{SV\,30}$	5.6	4.4	5.2	5.6	6.8
	Sludge volume index $I_{SV\,24}$	5	4.2	4.8	5.6	6.4
	Turbidity	10.80	9.68	29.87	54.81	62.72
611BC	Sludge volume index $I_{SV\,30}$	5.4	5.2	5.4	7	6
	Sludge volume index $I_{SV\,24}$	5.2	5	5.2	6.6	5.6
	Turbidity	7.65	15.82	43.93	57.79	64.51
Z-7692 + N-300	Sludge volume index $I_{SV\,30}$	5	5.4	4.8	4.8	6
	Sludge volume index $I_{SV\,24}$	4.6	5.2	4.4	4.4	5.6
	Turbidity	20.04	9.79	18.23	38.92	62.67
CF-2501 + N-300	Sludge volume index $I_{SV\,30}$	6	5.6	4.8	6	5.6
	Sludge volume index $I_{SV\,24}$	5.6	5.6	4.6	5.6	5.2
	Turbidity	31.87	23.89	7.27	19.56	39.69
611BC + N-300	Sludge volume index $I_{SV\,30}$	6.4	5.8	5.2	5.6	5.2
	Sludge volume index $I_{SV\,24}$	6.2	5.6	5	5.6	5
	Turbidity	55.65	16.83	17.20	27.34	47.72

The dual-additions was performed by pre-treating 250 mL of the working slurry in the flocculator by adding 50 mL of dilute solution of dual-polyelectrolytes (Z-7692+N-300, CF-2501+N-300, 611BC+N-300) of varying concentrations (0.1–0.8 g/L) in order to achieve a polymer combination ratio of 1:2. This combination was found to be more effective in preliminary tests performed prior to the experiments. The flash and slow mixing and subsequent measurements were carried out as described in section 4.2.2.

7.1.2 Particle-polymer interaction and selection of optimum dose

The structure of kaolin particle is comprised of pseudo-hexagonal platelets with differing surface chemistry occurring at the basal face and edge face (Hanson & Cleasby, 1990; Hocking et al., 1999; Lagaly, 1993, 2005). The basal face has a negative net surface charge that is usually unaffected by the pH. However, the charge on the edge face can be positive, neutral or negative depending on the pH of the suspension (Figure 4.2 and Figure 7.2). This behavior of kaolin platelets in aqueous suspension stems from proton gain or loss of surface hydroxyl groups and this strongly influences their flocculation properties (Goodwin, 2004; Sabah & Erkan, 2006).

The aggregation of dispersed kaolin particles in suspension can proceed with the polymer adsorbing either in an edge-to-face (EF) or face-to-face (FF) manner

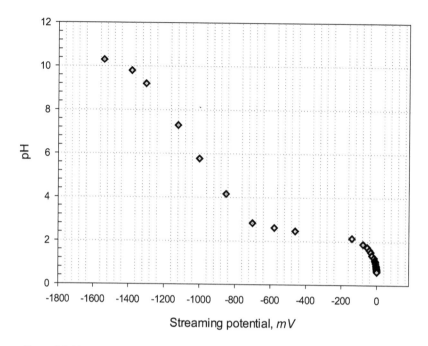

Figure 7.2 Variation of the kaolin slurry net surface charge with the pH (2% wt.).

depending on the prevailing flocculation conditions. At a pH lower than 7, edge-face electrostatic interaction predominates among the clay particles and a "card house structure" emerge. In this model, the mineral particles are held together by EF contacts (Goodwin, 2004). However, in the presence non-ionic polymer, the FF association is favored due to the larger surface area of the face compared to the edge of the kaolin particle (Kim & Palomino, 2009).

The kaolin slurry used in this study was conditioned at a pH of between 6 and 7 using different polymer combinations. The optimum polymer dose was thereafter determined based on the lowest observed values of the supernatant turbidity and sludge volume index I_{sv} after 30 mins and 24 hrs of settlement. However, choice of optimum flocculant dose might sometimes necessitate a compromise between the performance indicators (sludge volume index I_{sv} or supernatant quality). In the case of the chosen separation process (pelleting flocculation), it is assumed that in full-scale operations, residual water can be recycled back into the municipal sewer network. Hence the supernatant quality—clear water with low turbidity, is therefore a secondary criterion.

7.1.3   Polymer-dose response in single polymer treatment

Zetag® 7692, Sedipur® CF-2501 and Praestol® 611BC are medium to high molecular weight cationic polymers with low charge densities (Table 4.3). The choice of the polymer dose range reflects the regions of under-dose and over-dose (Al Momani & Örmeci, 2014). There was a significant reduction in the residual turbidity with all the polymer doses, and this improvement in supernatant clarity generally appears to deteriorate as the polymer dose increases with the exception of Sedipur® CF-2501. In the case of Sedipur® CF-2501, there was a slight reduction in turbidity when the polymer dose was increased from 1 to 2 kg/t TS and thereafter increases with the increase in dosage (Figure 7.3).

The sludge volume index I_{sv} after 30 mins of settling shows a somewhat similar trend with two clear regions of under-dose and over-dose of polymers and this is consistent with similar results reported elsewhere (Bache & Gregory, 2007; Bache & Zhao, 2001; Hemme et al., 1995). In the conditioning experiments, the optimum dosage of 2 kg/t TS roughly corresponds to the region of lowest turbidity and sludge volume index I_{sv} values. This can be deemed sufficient to obtain a good flocculation performance in terms of the two indicators.

In the under-dose range, increasing polymer dose improves the flocculation performance; however, once the optimum dosage condition is attained, additional polymer dose results in deterioration of flocculation efficiency. For instance, increasing the polymer dose from 1 kg/t TS to 2 kg/t TS leads to a decrease in the turbidity and sludge index to lowest values with the exception of Zetag® 7692 and Praestol® 611BC (Figure 7.3 and Figure 7.4) which show a slight increase in turbidity. Increasing the dosage beyond 2 kg/t TS shows a rapid increase in the observed values of the two indicators for all the polymers.

The response of the kaolin slurry to the physicochemical treatment can be explained using an appropriate conceptual model. The results of the experiments from this study confirm charge neutralization and polymer bridging as the main flocculation mechanisms with polymer bridging being more dominant. For high molecular weight polyelectrolytes with a charge density less than 15 %, the governing phenomenon is polymer bridging. Whereas for low molecular weight flocculant with ionicity

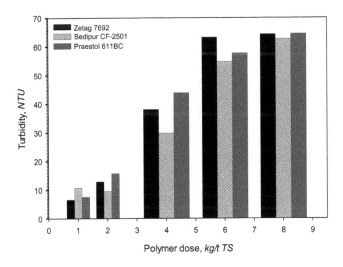

Figure 7.3 Relationship between polymer dosage and residual turbidity in single conditioning.

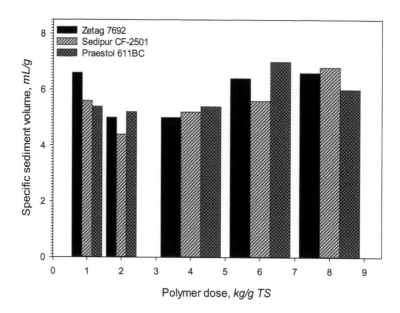

Figure 7.4 Relationship between polymer dosage and sediment volume in single conditioning.

greater than 30 %, charge neutralization predominates (Smith-Palmer & Pelton, 2006). The increase in turbidity at a higher dosage can be attributed to increasing polymer adsorption beyond the optimum which results in lower particle aggregation. The increasing coverage of the particle surface by the polymer molecules beyond the optimum level gradually leads to steric stabilization and a deterioration in the flocculation indicators (Smith-Palmer & Pelton, 2006).

Similarly, increasing polymer dose results in higher sludge volume index I_{sv} owing to the weakening of the polymer bonds and formation of more voluminous and bulky flocs (Smith-Palmer & Pelton, 2006). A theoretical charge reduction of approximately 48.4 %, 37.8 %, and 49.1 % for Zetag® 7692, Sedipur® CF-2501 and Praestol® 611BC respectively was obtained at the optimum dose of 2 kg/t TS. This confirms that optimum flocculation performance does not require a complete charge neutralization as illustrated in Figure 7.5 in the case of Z-7692 (Hjorth & Jørgensen, 2012).

7.1.4 Polymer-dose response in dual-polymer additions

Superfloc® N-300, a non-ionic medium molecular weight was used in combination with Zetag® 7692, Sedipur® CF-2501, and Praestol® 611BC in dual polymer conditioning. There are apparently two clear regions of under-dose and over-dose, with a generally good flocculation performance especially in turbidity reduction. This is however, achieved at a somewhat higher polymer dose (4 kg/t TS) when compared to the single polymer addition.

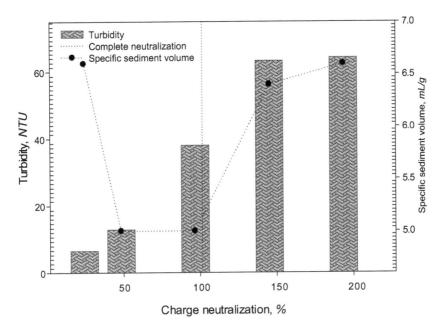

Figure 7.5 Surface charge neutralized by Z-7692 as a function of the flocculation parameters.

In terms of the sludge volume index I_{sv} and the residual turbidity, the dual addition results show that in the under-dose range, increasing the flocculant dose generally improves the flocculation performance except in the case of Z-7692+N300 (Figure 7.6 and Figure 7.7). Increasing the polymer dose beyond 4 kg/t TS however, shows a rapid deterioration in the flocculation efficiency as shown by the increase in the observed values of the two indicators. This region shows a significant increase in the turbidity and sludge volume index I_{sv} values indicating an over-dose of flocculants. These optima trends appear to be consistent with similar dosage optimization results reported elsewhere (Bache & Gregory, 2007; Bache & Zhao, 2001; Dentel, 2010; Herrington, Midmore, & Watts, 1993; Hjorth et al., 2008; Sher, Malik, & Liu, 2013).

The dual additions show a somewhat comparable flocculation efficiency albeit at twice the polymer dosage. For instance, CF-2501+N300 combination shows a lower turbidity at optimum dosage as compared to the single addition. In terms of the sludge volume index I_{sv}, the observed values in single polymer flocculation are generally lower. This might be attributed to the fact that in dual systems, larger and stronger flocs are formed (Lee & Liu, 2000). The two flocculation schemes implemented in this study generally show a good correlation in terms of the flocculation performance.

In addition, it does seem that the dual additions scheme produced stronger and shear-tolerant flocs as a result of the additional bridging effects, but this could not be directly verified in this study. However, further investigation in the vortex reactors showed that the dual addition gives more stability and strength to the flocs.

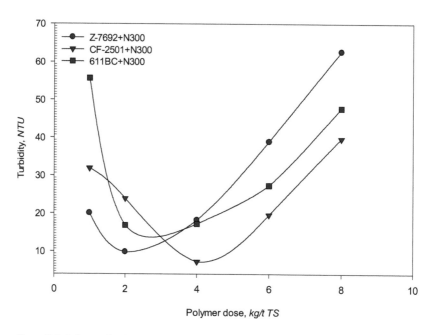

Figure 7.6 Relationship between polymer dose and residual turbidity in dual additions.

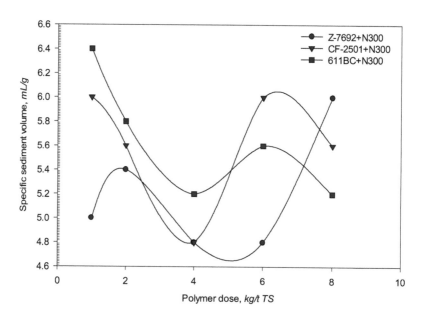

Figure 7.7 Relationship between polymer dose and sediment volume in dual additions.

Several studies have shown that the sequential addition of a cationic followed by a non-ionic polymer produces a synergistic effect that might result in stronger and more compact flocs due to the enhanced bridging effect of the non-ionic polymer (Ayol, Dentel, & Filibeli, 2005; Petzold, 1993). A conceptual model of the interaction mechanisms in dual-polymer flocculation was presented by Lee and Liu (Lee & Liu, 2000).

A theoretical charge reduction of approximately 31.30 %, 24.27 %, and 31.78 % for Z-7692+N-300, CF-2501+N-300, and 611BC+N-300 respectively was recorded at the optimum dose of 4 kg/t TS in dual-polymer combinations as shown for Z-7692+N300 combination in Figure 7.8. In contrast to the single conditioning, the optimum condition in dual additions was achieved at a slightly lower charge neutralization. The might be attributed to the negative charge of non-ionic polyacrylamide (N300) due its hydrolysis in solution.

7.1.5 Supernatant clarity at the optimum dosage

The effect of polymer characteristics, especially molecular weight, is a critical factor in the flocculation of kaolin slurry. It has been shown that polymer bridging is the dominant flocculation mechanism in the case of non-ionic or low-charge cationic polymers (Kim & Palomino, 2009). Polymer molecules can adsorb onto both the face and edge sites of the kaolin particle with face-to-face (FF) association more probable in the case of high molecular weight flocculants (Kim & Palomino, 2009).

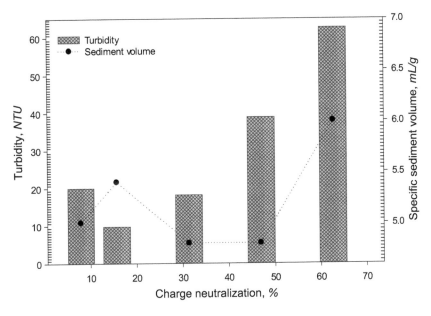

Figure 7.8 Surface charge neutralized by Z-7692+N300 combination as a function of the flocculation parameters.

Figure 7.9 shows the effect of polymer types on the supernatant turbidity at the respective optimum dose. The lowest turbidity of 7.27 NTU was recorded with CF-2501+N-300 polymer combination in the dual additions. It does appear that the dual-polymers have a slightly superior overall turbidity reduction over the single addition, although at a much higher optimum dosage.

In terms of the individual flocculants, Sedipur® CF-2501 gives the highest turbidity reduction with an observed value of 9.68 NTU. This might probably be attributed to its slightly higher molecular weight when compared to other flocculants, as it has been shown that this performance criteria is a function of the polymer molecular weight (Sabah & Erkan, 2006). A similar trend can be seen in its combination with Superfloc® N-300, which gives the lowest observed turbidity value for the dual-polymers. It appears that this trend seems to be consistent with all the other polymer combinations with Z-7692 and Z-7692+N-300 giving the second lowest values in single and dual additions schemes respectively.

7.1.6   Specific sediment volume at optimum dosage

The characteristics of the settled flocs is an important factor in many solid-liquid separation processes as the flocs' structure and appearance is an important indicator of flocculation efficiency (Smith-Palmer & Pelton, 2006). The aim of pelleting flocculation as a pre-treatment technique is to produce compact pellet flocs with high solids content, shear resistance, and better dewaterability. The sludge volume is a simple and effective indicator to assess the settleability characteristics of sludge and the flocculation performance. Figure 7.10 shows the effect of flocculant types on the sludge volume index I_{SV} at the respective optimum conditions.

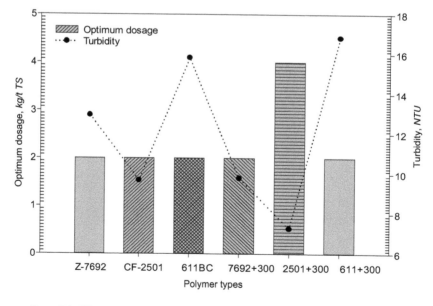

Figure 7.9 Effect of polymer types on supernatant clarity at the optimum dosage.

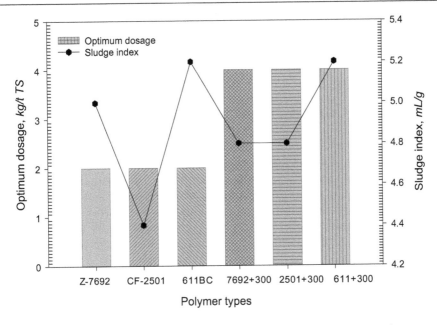

Figure 7.10 Effect of polymer types on sediment compaction at the optimum dosage.

The lowest observed sludge index value of 4.4 mL/g was obtained with Sedipur® CF-2501 in the single polymer conditioning. This might be as a result of the flocculant's higher molecular weight as polymer bridging is a function of molecular weight in the case of low charge cationic polymers. In the case of dual-additions, the lowest observed value is 4.8 mL/g for both Z-7692+N-300 and CF-2501+N-300 polymer combinations. This slightly higher value obtained at a higher optimum dose indicates that the dual-additions produce less compact flocs. However, the shear resistance of the flocs, which is another critical factor in the pelleting process, was not directly measured in the physicochemical study.

7.1.7   Process relevance of the physicochemical parameters

In practice, the variations observed in the values of the measured parameters can be directly related to pelleting process in the choice of the optimum dose. A trade off can be made when evaluating the significance of each parameters. For instance, nearly all the flocculants tested achieve an acceptable turbidity reduction with very low residual particles in suspension even at low polymer dosage. Consequently, the significance of the performance improvement of this parameter is very minimal, hence it cannot be used to justify higher cost of conditioning chemicals (i.e. higher polymer does). Therefore, the choice of CF-2501+N300 at twice the polymer dose cannot be justified by the marginal improvement in residual turbidity.

In contrast to the residual turbidity, the specific sediment volume exhibits a significant improvement at higher polymer dose. In practical terms, this means that the flocs' compactness and solids content will increase appreciably with higher polymer dose and this will be advantageous in terms of cost savings in the subsequent sol-id-liquid separation. Bulky and less compact flocs tend to cause operational problems because they are prone to breakage and slow to settle compared to more compact pellet flocs.

7.2   OPTIMIZATION OF THE PELLETING PROCESS

The agglomeration process was randomly optimized for a number of process vari-ables and reference points using quasi-factorial experimental design to select the optimization points (Table 7.2). Several batch flocculation and gravity dewatering experiments were performed over a range of different process conditions to deter-mine a number of optimum reference points in which the process conditions (i.e. physicochemical and micro-hydrodynamics) will result in the formation of pel-let-like aggregates. A critical mixing speed of 170 rpm and 195 rpm was observed in the batch and full-volume continuous reactors respectively, above which there was a significant rupture of the pellet flocs and this trend appears to intensify with the mixing speed. In addition, the flow pattern and the effect of the operating con-ditions on the process performance were also investigated. The preparation and characterization of the agglomerates and the reactor supernatant were carried as described sections 4.2.6 and 4.2.4.

The optimization of the micro processes that influences the pre-treatment tech-nique was performed in the flow devices with different geometries. The results of agglomeration and gravity dewatering experiments and the corresponding optimum process conditions for the pelleting process are described in the following sections. A summary of the results of the effect of micro-processes on the pelleting process in the batch reactor are presented in Table 7.3.

7.2.1   Effects of micro processes on the pelleting process

The influence of the micro-processes on the pelleting process in the batch reactor was investigated by assessing the clarity of the reactor supernatant and the particle removal efficiency R_x after gravity dewatering at intervals of 5, 10, 15 and 20 minutes.

Table 7.2 Process operating parameters and reference points for the batch experiments

Operational parameters	Reference points			
Polymer type	Cationic	Anionic	Cationic-Non-ionic	Anionic-Non-ionic
Polymer dose, kg/t TS	2	3	4	5
Dosing sequence	Once	Twice	Thrice	
Slurry concentration, % wt.	1	2	3	4
Filling mode	Batch	Quasi-continuous	Fully-continuous	
Wall-plate gap, mm	3	4	5	6
Mixing speed, min⁻¹	125	145	165	185

Table 7.3 Summary of indicators of process performance at optimum conditions for the batch process

Polymer dose C-492+N-300) (kg/t TS)	Indicators of process efficiency	Residence time (s)			
		300	600	900	1200
3 (1+2)$_{145\ rpm}$	Turbidity (NTU)	48.872	38.876	25.277	30.348
	Removal efficiency (%)	95.625	94.375	84.375	90
	Solids content (%)	24.9592	24.7541	27.7207	26.4220
4 (1.5+2.5)$_{145\ rpm}$	Turbidity (NTU)	66.246	39.322	38.729	27.108
	Removal efficiency (%)	96.25	94.375	90.625	96.875
	Solids content (%)	26.1905	24.3156	26.3636	26.8631
5 (2+3)$_{145\ rpm}$	Turbidity (NTU)	98.965	118.81	61.778	33.101
	Removal efficiency (%)	97.50	92.50	93.75	91.875
	Solids content (%)	26.2185	26.2411	26.8817	25.3448
3 (1+2)$_{165\ rpm}$	Turbidity (NTU)	36.401	25.288	14.513	7.972
	Removal efficiency (%)	90.625	94.375	85.625	74.375
	Solids content (%)	27.7247	27.5547	28.3644	27.9343
4 (1.5+2.5)$_{165\ rpm}$	Turbidity (NTU)	61.622	47.144	25.759	29.655
	Removal efficiency (%)	96.875	95.625	96.25	93.125
	Solids content (%)	26.9505	27.0318	27.0650	27.2395
5 (2+3)$_{165\ rpm}$	Turbidity (NTU)	86.417	64.269	45.409	69.655
	Removal efficiency (%)	97.5	95	89.375	91.25
	Solids content (%)	25.5738	27.2890	26.6294	27.3408

The results of the residual turbidity show a gradual decrease with residence time and agitation speed, whereas the residual turbidity increases with polymer dosage above the optimum dosage of 3 kg/t TS. The lowest turbidity of 25.3 and 7.9 NTU were obtained with a polymer dose of 3 kg/t TS at both agitation speeds of 145 and 165 rpm respectively (Figure 7.11 and Figure 7.12). The optimum polymer dose of 3 kg/t TS shows a good quantitative agreement with the physicochemical optimization (Oyegbile, Ay, & Narra, 2016).

The observed trend in the turbidity measurements with respect to residence time can be attributed to an increased floc attachment and growth that allows micro-particles to be incorporated onto the macro-flocs, which in turn results in lower turbidity. With respect to the agitation speed, the observed reduction in turbidity can be attributed to the increased flocculation kinetics and inter-particle collisions. It has been reported that increases in agitation speed also increase the shear rate, which promotes particle-particle contacts and floc growth up to the steady state condition and this ultimately results in improved particle aggregation and lower turbidity (McConnachie, 1991; Rulyov, 2010; Spicer & Pratsinis, 1996). However, in terms of the polymer dosage, a higher polymer dose appears to correlate well with an increase in turbidity of the supernatant.

The increase in turbidity at a higher dosage can be attributed to unadsorbed polymer molecules that remain in solution due to increasing polymer adsorption beyond

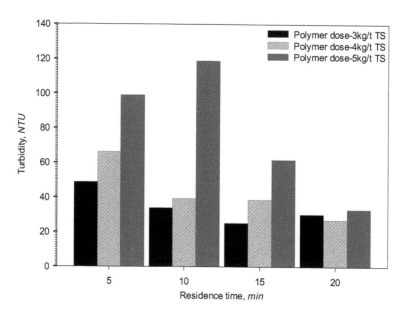

Figure 7.11 Relationship between residence time and supernatant quality at operating speed of 145 rpm.

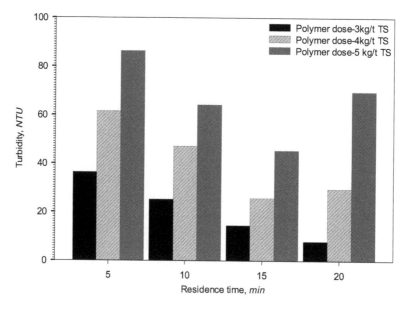

Figure 7.12 Relationship between residence time and supernatant quality at operating speed of 165 rpm.

the optimum. This results in lower particle aggregation due to the weakening of the polymer bond (Besra et al., 2002; Betatache et al., 2014).

The observed values of the removal efficiency R_x generally show a gradual decrease with residence time and operating speed except at a longer residence time (>15 minutes) when the efficiency shows a slight increase (Figure 7.13 and Figure 7.14). In these cases, the removal efficiency shows a gradual increase at a residence time of 20 minutes. This observation might be attributed to the fact that in both cases the compact agglomerate undergoes a gradual deformation with time as the floc size increases and becomes more susceptible to breakage as shown by the increase in the proportion of fines (< 0.5mm) in the reactor system (He et al., 2015; Wu & Patterson, 1989; Yeung & Pelton, 1996).

While this trend does not seem to correlate with the trend in the turbidity measurements, the observed variation might be attributed to the fact that the eroded micro-particles were not effectively captured by the sieve, resulting in a lower than expected removal efficiency. However, in terms of the agitation speed, there was no significant differences in the observed values of the particle removal efficiency.

The residual charge in the supernatant remains negative irrespective of the operating speed and polymer dose. This negative charge is due to the partial theoretical charge neutralisation of approximately 35.6 %, 52.5 %, and 69.4 % for polymer doses of 3, 4, and 5 kg /t TS respectively. This confirms the theory that an optimum flocculation performance does not entail a complete charge reversal even in the case of dual polymer additions (Lee & Liu, 2000).

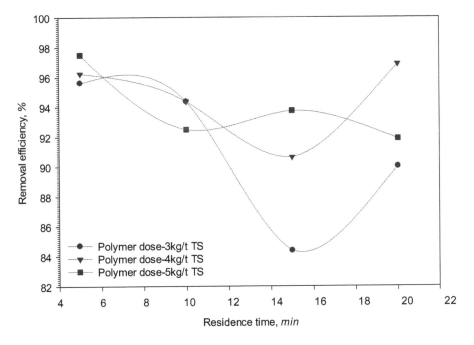

Figure 7.13 Relationship between residence time and the particle removal at operating speed of 145 rpm.

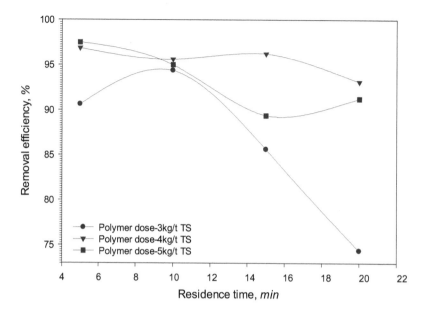

Figure 7.14 Relationship between residence time and the particle removal at operating speed of 165 rpm.

7.2.2   Effect of optimum dose on the process efficiency

In terms of the observed parameters, the process performance at the optimum dose of 3 kg/t TS showed the lowest residual turbidity values (7.972 NTU) while a dosage of 5 kg/t TS recorded the highest particle removal efficiency (97.5 %). Therefore, in full-scale operations, results of the residual turbidity clearly demonstrate that the residual water can be safely utilized for non-portable reuse or recycled back into the municipal sewer network. By contrast, the difference in the removal efficiencies at the respective dosages is very minimal (~1.875 %). Hence, in practical applications, this performance improvement will be insignificant and hence will not justify a higher cost of conditioning chemicals.

7.2.3   Effects of micro processes on the solids fraction

The effect of the micro-processes on the subsequent phase separation efficiency by simple gravity dewatering was investigated by determining the dry solids content of the dewatered pellets. The results of the agglomerates' solids content at residence time intervals of 5, 10, 15 and 20 minutes are presented in Figure 7.15.

The average dry solids content of the agglomerates does not appear to show any significant variation with respect to the residence time once the pellet flocs are fully formed after 5 minutes. However, in terms of the agitation speed, a slightly higher solids content was observed for the pellets obtained at a rotation speed of 165 rpm. The highest solids content of 28.3 % was recorded after 15 minutes in all experimental conditions. In addition, the highest solids content of 27.7 % and 28.3 %

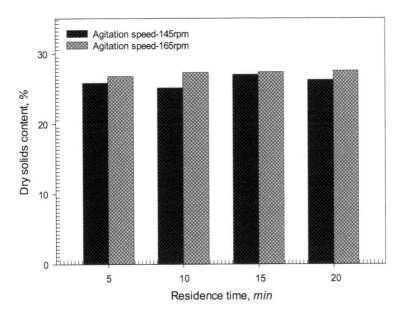

Figure 7.15 Relationship between residence time and the dry solids content at different operating speed.

were obtained at a polymer dose of 3 kg/t TS and agitation speeds of 145 rpm and 165 rpm respectively. This seems to confirm a polymer dose of 3 kg/t TS as the optimum (Lee & Liu, 2000).

In contrast, there seems to be no correlation between solids content and removal efficiency R_x; in fact, contrary to what might be expected at a lower solids content, there is a higher removal efficiency. One possible explanation for this might be that the pellet flocs with higher removal efficiency contains micro particles that were not effectively captured in the gravity dewatering thereby giving a lower than expected dry solids content values.

The observed process efficiency for this pre-treatment technique appears to compare quite favourably with conventional solid-liquid separation techniques reported in literature (Carissimi & Rubio, 2015; Mahmoud et al., 2013). In addition, the optimum polymer dosage value of 3kg/t TS obtained in this study appears to be much lower than those reported in similar studies, although a different substrate was employed (Sievers et al., 2008).

7.2.4 Predicted and effective polymer dose

The aim of the physicochemical optimization is to select the appropriate flocculant and dosage or range of dosage. In order to accurately predict the polymer dose, a simulation of the physicochemical parameters under conditions that are similar to that of the pelleting process was performed in the jar flocculator. The optimum dosage values of 2–4 kg/t TS obtained in the physicochemical investigation shows fairly good correlation with the optimum dosage in the pre-treatment process (3 kg/t TS). The physicochemical optimization has been successfully employed for effective polymer dose prediction for the pelleting process. However, it should be pointed out that a more detailed analysis of the intrinsic physicochemical properties of the substrates might be used to further optimize the pelleting process.

7.3 PROCESS RELEVANT PELLET CHARACTERISTICS

The effect of the micro-processes on the subsequent phase separation efficiency and the stable pellets' properties was structural investigated by determining the process relevant parameters such as dry solids content, shape, size, and size distribution, as well as the structural stability (compressive strength) of the agglomerates.

7.3.1 Image and particle size analysis

Samples of the dewatered pellets were observed under digital microscope (Keyence VHX-2000) at high resolution in order to determine the size, shape and distribution of the samples. The modular system comprises of replaceable lens, camera and a controller unit for image acquisition and processing. The results of the image and particle size analysis of dewatered pellets show that the pellets exhibit a narrow size distribution with a nearly spherical shape as illustrated in Figure 7.16. A mean equivalent pellet size of 3.7582 mm and 3.8460 mm for rotation speed of 145 and 165 rpm respectively were obtained from the pelleting process.

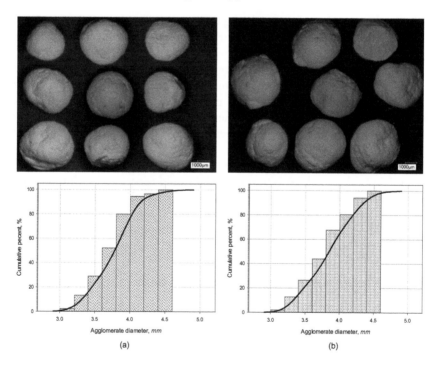

Figure 7.16 Microscopic image and size distribution of dewatered pellets obtained from the batch reactor (a) at 145 rpm (b) at 165 rpm.

The diameter of the dewatered pellets from the batch process shows a narrow size distribution. Pellet size ranges from 3.085 to 4.628 mm at operating speed of 145 rpm and 3.145 to 4.71 mm at a speed of 165 rpm. This minor difference could be attributed to the increased floc growth at higher shear rate and hydrodynamic force below the critical floc size.

7.3.2   Agglomerate strength analysis

Test agglomerates were compressed at an axial velocity of 0.1667 mms^{-1} until they were observed to fracture with the maximum force at fracture point recorded on a force-displacement chart. The results of the compressive strength as a function of the equivalent pellet diameter and the force-displacement data are presented in Figure 7.17 and Table 7.4. In addition to the agglomerate compressive strength, these results also provides information on the resistance of the pellets to abrasion as it has been reported that the pellet's resistance to attrition is closely related to the agglomerate compressive strength (Barbosa-Canovas et al., 2005).

The result clearly demonstrates that the mean agglomerate compressive strength decreases with the pellet diameter for both agitation speed. There was a distinct first maximum for most of the pellets in the force-displacement data that indicate the

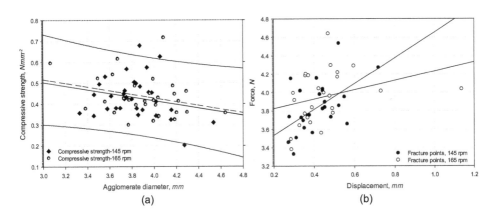

(a)   (b)

Figure 7.17 Compressive strength of dewatered pellets a function of the diameter and the strain rate for single point fracture (a) compressive strength (b) strain rate.

Table 7.4 Pellet characteristic test properties for the compression test

	Agitation speed	
Test parameters	145 rpm	165 rpm
Mean pellet diameter, mm	3.8330	3.9182
Mean compressive strength, Nmm^{-2}	0.4298	0.4351
Mean strain rate, s^{-1}	0.3639	0.4088
Mean maximum compressive force, N	4.9476	5.2303

precise point at which the agglomerate was observed to fracture. Few of the test samples recorded multiple fracture points and the maximum force was recorded at the first fracture point. In addition, the plot of the test parameters (compressive strength and force-displacement) as a function of the pellet size shows a narrow distribution of these parameters with their observed values decreasing with increase in pellet size. In terms of the agitation speed, there is a slight increase in the mean compressive strength at higher rotation speed which seem to suggest that most of the pellets are of comparable strength. Nevertheless, it can be argued that increased rotation speed improves the structural stability of the pellets due to increased flocs' adhesion.

In the final analysis, the main objective of the pelleting process which is to produce compact aggregates with uniform characteristics (i.e. shape, size, structural stability etc.) has been achieved with the choice of appropriate flow units and process conditions. However, higher process performance can be achieved by further optimization of the flow units and the operating conditions.

Chapter 8

Conclusions and perspectives

This study has been conducted to demonstrate the structure formation process in a novel pre-treatment technique as a technical proof of concept. The simulation results show that the optimum process condition is a function of the physicochemical parameters (suspension particle size and concentration, agitation time, polymer type and dosage, charge density and addition rate etc.) process engineering (residence time T, intensity and duration of shear, stirrer-vessel configuration etc.), and hydrodynamic conditions in the simulation vessels. In the final assessment of the pelleting process, it is important to emphasize the influence of the reactor's rotor-stator configuration and the operating conditions on the pelleting process.

8.1 EFFECT OF ROTOR-STATOR CONFIGURATION

The influence stirrer-vessel configuration on the pelleting process can be quite significant. In fact, the realization of the pelleting technique is predicated upon the right choice of reactor geometry and mixer. The two reactors tested in the course of this investigation differs slightly in terms of their geometry while the stirrer configuration is similar. In spite of this, the results of this study show a certain degree of differences in terms of the power consumption, flow pattern and vortex structure. The continuous reactor recorded a higher power consumption at the same operating speed compared to the batch reactor but could be operated at a much lower agitation speed (~70 rpm) thereby reducing its power consumption. In addition, the continuous reactor shows a more uniformly distributed velocity and vorticity pattern.

The data from this study clearly shows that the flow streamline and vortex pattern is a function of the vessel geometry. The geometry of the continuous reactor provides a superior hydrodynamic condition for the pelleting process in terms of the averaged velocity vector and vorticity magnitude and their distribution within the flow field as compared to the batch reactor. In addition, there is a variation in the flow pattern along the rotor-stator cavity with an identical flow pattern near the rotor and at the center of the cavity. However, due to high level of instability in the flow, the flow measurement could not be performed at all the operating speeds for the pelleting process.

In terms of the rotor configuration, the results of the hydrodynamic analysis show that the stirrer configuration influences the mixing and the power consumption in both reactors significantly. The continuous reactor shows a higher kinetic energy dissipation rate and consequently higher mass transfer rate and energy consumption compared to the batch reactor. The flow in both reactors are fully turbulent under the operating conditions while the power curve for the mixer shows a good similarity with common types of agitators found in published literature (Godfrey & Amirtharajah, 1991; Thoenes, 1998).

8.2 EFFECT OF PROCESS CONDITIONS

The results of this study clearly show that the residence time T and shear regime G strongly influence the pelleting process. Increasing shear rates and residence time ultimately results in floc degradation and disruption. An equilibrium point for both parameters is therefore required to significantly improve the pelleting process. However, in terms of shear tolerance, the continuous reactor was found to provide a more shear-tolerant environment than the batch reactor by virtue of its geometry.

The choice of appropriate flocculant and dosage has a profound effect not only on the pelleting process but also the cost of pre-treatment. The result of the physicochemical optimization in this study shows that molecular weight of polymer is an important factor in polymer selection. An ideal polymer for this process must provide maximum bridging effects at low dosage. This study also demonstrated that it is also possible to carefully select a combination of polymers in dual-conditioning in order to take advantage of their synergistic effect.

8.3 FUTURE PERSPECTIVES

This study presents an evaluation of the effects of micro hydrodynamics and physicochemical micro processes on the aggregation of model substrates using an innovative shear-assisted pre-treatment technique. Results of the investigation showed that the effectiveness of the aggregation process is highly dependent on an optimum combination of several process variables such as the polymer type, concentration and dosing regimen, agitator rotation speed, wall-plate gap distance, residence time, suspension loading method, reactor geometry and stirrer type. While the optimization of the micro processes has been carried out as far as possible within the time frame of this study, the effects of other process parameters that were not fully considered in the course of this research work cannot be totally ignored. It is anticipated that future studies will address the flow optimization especially the realization of a fully continuous laboratory flow unit and the scale-up of the reactor systems.

8.3.1 Effect of other process conditions

There are several operating parameters that influence the micro processes, a number of which were optimized for the reactor systems in this study. However, an optimum combination of several operating variables under laboratory conditions is not only tedious but time consuming. In future studies, it is envisaged that the effect of several other operating parameters such as slurry concentration, stirrer configuration, structure of the vortex field etc., which were not fully considered in this study will be further investigated experimentally. It is hoped that in conducting such studies, a combination of experimental and computational techniques can be employed in order to provide a more detailed analysis of the pre-treatment process.

8.3.2   Computational study of the fluid flow

The analysis of the nature and structure of the flow stream pattern has been performed using experimental technique (PIV). However, due to high degree of flow instability at higher rotation speed, there was a high degree of fluctuations in the laser signals which makes measurements under such conditions quite challenging.

The use of computational techniques in optimizing the flow conditions and flocculation process in flow units has been fast gaining wider acceptance as a research tool. It will allow the hydrodynamics of the pelleting process to be simulated by taking into account the flow patterns and other parameters such as the velocity, turbulent kinetic energy dissipation, and local velocity gradient. Therefore, it is anticipated that in the future, a more comprehensive and detailed analysis of the turbulent flow field in the vortex reactors can be performed using CFD. Future investigation using computational fluid dynamics will provide more information on the nature of the flow field and its effect on the pelletization process. In addition, optimization of the laboratory flow units can be conveniently performed within the computational environment by varying the different operating parameters influencing the fluid flow such as the stirrer-vessel configurations (Bridgeman, 2008, 2010; Craig et al., 2002; Das et al., 2016; Essemiani & De Traversay, 2002; Korpijärvi et al., 2000; Samaras et al., 2010; Torfs, 2015). This will save time and significantly reduce the cost of future investigations. In turn, the computational results can be compared to the experimental data to establish the accuracy of the models and for validation purposes.

8.3.3   Numerical modeling of the pelleting process

The numerical modelling of the structure formation and phase separation processes (wet agglomeration) by pelleting aggregation technique will provide information on the time-evolution of the floc structure in the turbulent flow field (Bridgeman et al., 2010; Hessel et al., 2004; Lee & Molz, 2014; Marchisio, Vigil, & Fox, 2003; Wang et al., 2005). While there are several studies on the modeling of classical orthokinetic flocculation, there are very few numerical studies on the pelleting flocculation process. Future studies combining experimental data with the mathematical modeling (PBM) of pelleting flocculation will provide much-needed information for an improved process control, and will simplify the design, optimization and scale-up of pelleting reactors for multi-phase separation (Heath & Koh, 2003; Kharoua, Khezzar, & Saadawi, 2013; Nopens, 2007; Oshinowo, Elsaadawy, & Vilagines, 2015; Oshinowo, Quintero, & Vilagines, 2016; Prat & Ducoste, 2007; Prat & Ducoste, 2006; Torfs, Vesvikar, & Nopens, 2013).

8.4   CONCLUDING REMARKS

Process optimization has been shown to be a crucial step in the design and development of process equipment. Results from this research work demonstrate that an optimum combination of several operating parameters results in improved process performance. It is anticipated that future studies will address the suggested improvements, and the scale-up of the pre-treatment units.

Appendix

A.1 FIGURES

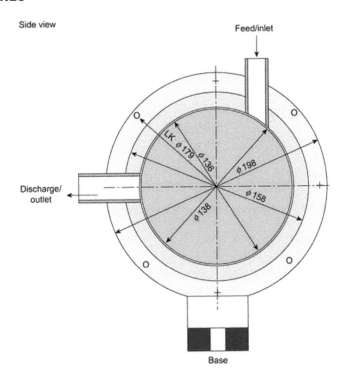

Figure A.1 Side view of the continuous vortex reactor (not to scale, all dimensions in millimeters).

Side view with details

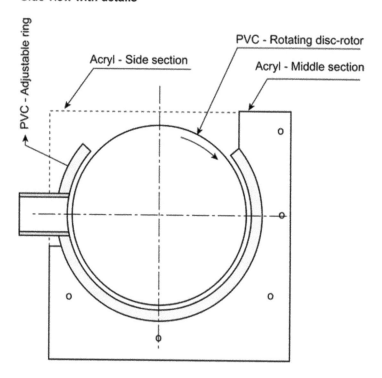

Figure A.2 Vertical cross section of the continuous vortex reactor A—A.

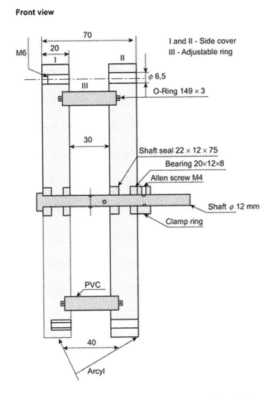

Figure A.3 Vertical cross section of the continuous vortex reactor B—B (not to scale, all dimensions in millimeters).

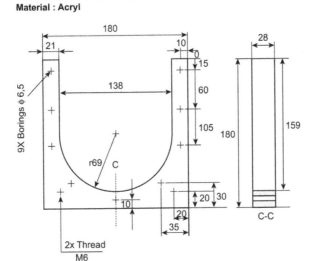

Figure A.4 Side view and cross section C—C of the batch vortex reactor (not to scale, all dimensions in millimeters).

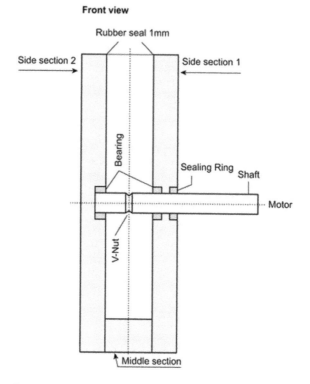

Figure A.5 Vertical cross section of the batch vortex reactor.

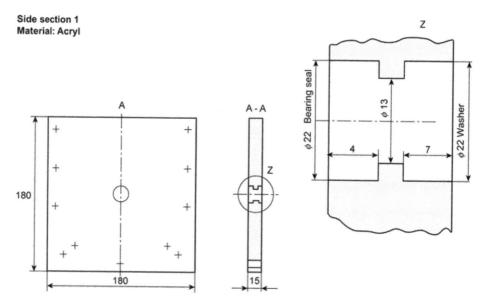

Figure A.6 Cross sectional view at the rotor of the reactor A—A (not to scale, all dimensions in millimeters).

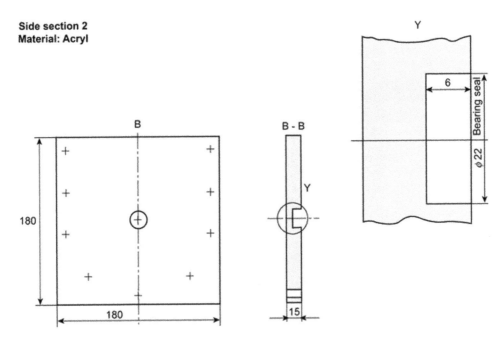

Side section 2
Material: Acryl

Figure A.7 Cross sectional view at the stator of the reactor B—B (not to scale, all dimensions in millimeters).

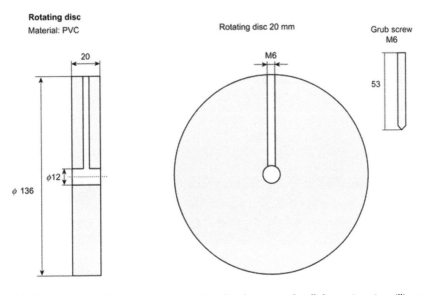

Rotating disc
Material: PVC

Figure A.8 Cross section of the agitator—rotating disc (not to scale, all dimensions in millimeters).

Figure A.9 Pellet flocs from kaolin slurry in continuous reactor, 3 kg/t TS, 180 rpm, 20 min.

Figure A.10 Pellet flocs from Ferric hydroxide slurry in continuous reactor, 5 kg/t TS, 180 rpm, 20 min.

Figure A.11 Pellet flocs from kaolin slurry in batch reactor, 3 kg/t TS, 145 rpm, 20 min.

Figure A.12 Wet pellet flocs from kaolin slurry in batch reactor after gravity dewatering, 3 kg/t TS, 145 rpm, 20 min.

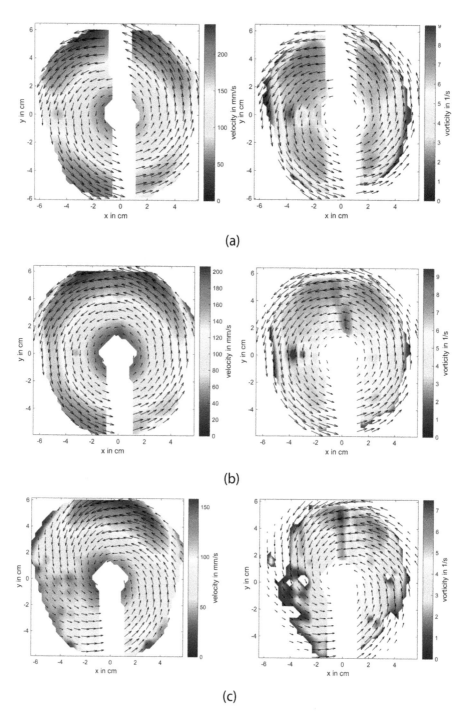

Figure A.13 2-D flow stream and vortex pattern in continuous reactor at 90 rpm showing the averaged velocity vector field, mm s⁻¹ and vorticity map, s⁻¹ (a) near the rotor (b) at the center of cavity (c) near the stator.

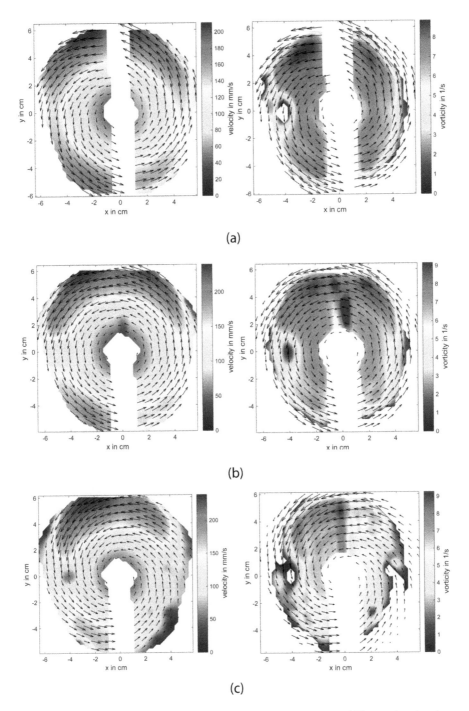

Figure A.14 2-D flow stream and vortex pattern in continuous reactor at 110 rpm showing the averaged velocity vector field, mm s^{-1} and vorticity map, s^{-1} (a) near the rotor (b) at the center of cavity (b) near the stator.

A.2 TABLES

Table A.1 Characteristic diameters of dewatered kaolin pellets from batch reactor at a rotation speed of 145 rpm

Pellet Nr.	d_1 (mm)	d_2 (mm)	d_{av} (mm)	d_{eq} (mm)
1	3.365	3.217	3.291	3.202
2	4.184	3.484	3.834	3.925
3	4.610	4.010	4.310	4.445
4	4.088	3.780	3.934	3.993
5	4.014	3.803	3.909	3.803
6	3.834	3.398	3.587	3.722
7	3.663	3.322	3.493	3.471
8	3.431	3.376	3.404	3.417
9	4.174	3.909	3.542	4.118
10	3.642	3.098	3.370	3.267
11	3.876	3.078	3.477	3.681
12	3.769	3.492	3.631	3.547
13	4.035	3.811	3.923	4.019
14	3.845	3.588	3.717	3.728
15	3.716	3.332	3.524	3.469
16	3.888	3.866	3.877	3.896
17	3.386	3.291	3.339	3.305
18	4.930	4.344	4.637	4.628
19	4.208	4.205	4.207	4.117
20	3.641	3.546	3.594	3.582
21	4.024	3.897	3.961	3.924
22	3.929	3.450	3.690	3.705
23	4.196	3.907	4.052	3.960
24	3.843	3.535	3.689	3.732
25	4.089	3.914	4.002	4.014
26	3.952	3.918	3.933	3.915
27	3.918	3.696	3.807	3.805
28	4.354	3.995	4.175	4.182
29	4.133	3.396	3.765	3.996
30	3.654	3.184	3.419	3.612
31	3.817	3.727	3.772	3.810
32	4.472	4.328	4.400	4.446
33	3.976	3.706	3.841	3.857
34	3.758	3.648	3.703	3.746
35	4.237	3.739	3.988	3.775
36	3.398	3.151	3.275	3.250
37	3.739	3.513	3.626	3.714
38	4.068	3.537	3.803	3.789
39	3.886	3.642	3.764	3.638
40	4.259	3.876	4.068	4.094
41	4.048	3.973	4.011	3.989
42	3.325	3.304	3.315	3.336
43	4.036	3.428	3.732	3.651
44	3.780	3.228	3.504	3.483
45	3.514	3.430	3.472	3.473
46	3.311	3.216	3.264	3.325
47	4.004	3.822	3.913	3.865

(Continued)

Table A.1 (Continued).

Pellet Nr.	d_1 (mm)	d_2 (mm)	d_{av} (mm)	d_{eq} (mm)
48	3.801	3.609	3.705	3.744
49	4.274	3.229	3.752	3.939
50	3.430	3.408	3.419	3.437
51	4.133	3.673	3.903	3.980
52	3.750	3.205	3.478	3.391
53	3.920	3.790	3.855	3.839
54	4.355	4.134	4.245	4.257
55	4.014	3.259	3.637	3.546
56	4.248	3.983	4.116	4.117
57	4.408	3.737	4.073	4.027
58	4.504	3.909	4.207	4.107
59	3.717	3.397	3.557	3.525
60	3.995	3.833	3.914	3.886
61	4.440	4.286	4.363	4.345
62	3.897	3.397	3.647	3.687
63	4.173	3.909	4.041	3.928
64	3.951	3.590	3.771	3.795
65	3.918	3.791	3.855	3.906
66	3.631	3.408	3.520	3.640
67	3.379	3.311	3.345	3.403
68	3.955	3.652	3.804	3.775
69	3.962	3.782	3.872	3.960
70	3.929	3.461	3.695	3.476
71	3.503	3.366	3.435	3.414
72	3.748	3.453	3.601	3.603
73	4.291	3.783	4.037	3.877
74	3.493	3.002	3.248	3.208
75	4.461	3.396	3.929	3.769
76	3.291	3.232	3.262	3.273
77	4.067	3.941	4.004	3.964
78	4.132	3.993	4.063	4.106
79	4.250	3.962	4.106	4.066
80	3.940	3.663	3.802	3.863
81	3.922	3.769	3.846	3.909
82	3.822	3.653	3.738	3.663
83	3.620	3.185	3.403	3.394
84	3.535	3.269	3.402	3.414
85	3.120	2.951	3.536	3.085
86	3.579	3.439	3.509	3.642
87	4.090	3.887	3.989	3.939
88	3.197	2.864	3.031	3.113
89	4.169	3.822	3.996	4.049
90	4.301	4.017	4.149	4.043

Table A.2 Characteristic diameters of dewatered kaolin pellets from batch reactor at a rotation speed of 165 rpm

Pellet Nr.	d_1 (mm)	d_2 (mm)	d_{av} (mm)	d_{eq} (mm)
1	4.270	4.227	4.249	4.228
2	3.720	3.547	3.634	3.700
3	3.791	3.673	3.732	3.871
4	4.856	3.976	4.416	4.304
5	4.174	4.102	4.138	4.411
6	3.495	3.227	3.361	3.451
7	4.035	3.995	4.015	4.197
8	3.898	3.822	3.860	3.832
9	4.136	3.898	4.017	3.845
10	4.164	3.717	3.941	3.977
11	4.770	4.356	4.563	4.461
12	4.005	3.918	3.962	3.921
13	3.901	3.556	3.729	3.594
14	4.280	4.113	4.197	4.150
15	4.356	3.610	3.983	4.244
16	3.483	3.312	3.398	3.274
17	4.419	4.215	4.317	4.386
18	4.302	4.273	4.288	4.292
19	4.163	3.875	4.218	4.112
20	4.467	4.003	4.235	4.205
21	3.654	3.621	3.638	3.832
22	3.675	3.599	3.637	3.775
23	3.886	3.537	3.712	3.747
24	4.567	4.498	4.533	4.370
25	5.272	3.728	4.500	4.709
26	4.440	4.282	4.361	4.378
27	4.099	3.824	3.962	4.081
28	4.174	3.706	3.940	3.988
29	3.738	3.312	3.525	3.642
30	3.950	3.444	3.697	3.560
31	3.711	3.706	3.709	3.831
32	3.599	3.548	3.574	3.431
33	4.046	3.854	3.950	3.708
34	4.105	4.057	4.081	4.124
35	3.961	3.687	3.824	3.899
36	4.632	3.900	4.266	4.465
37	3.811	3.443	3.627	3.440
38	3.825	3.631	3.728	4.232
39	4.587	4.111	4.349	4.299
40	4.376	4.058	4.217	4.296
41	3.825	3.439	3.632	3.974
42	4.142	4.047	4.095	4.113
43	3.644	3.634	3.639	3.768
44	3.537	3.333	3.435	3.348
45	3.812	3.542	3.677	3.914
46	3.208	3.108	3.158	3.193
47	5.051	4.121	4.586	4.263
48	3.974	3.558	3.766	3.868
49	3.780	3.529	3.655	3.797

(Continued)

Table A.2 (Continued).

Pellet Nr.	d_1 (mm)	d_2 (mm)	d_{av} (mm)	d_{eq} (mm)
50	4.365	4.210	4.288	4.247
51	4.148	3.348	3.748	3.490
52	3.557	3.323	3.44	3.283
53	3.619	3.567	3.593	3.422
54	4.191	4.027	4.109	3.849
55	3.880	3.750	3.815	3.691
56	3.801	3.583	3.692	3.408
57	4.717	3.556	4.137	4.151
58	3.580	3.194	3.387	3.607
59	3.798	3.650	3.724	3.768
60	4.410	4.025	4.218	4.528
61	4.336	4.108	4.222	4.007
62	4.141	3.919	4.030	3.802
63	4.166	3.754	3.960	3.899
64	4.581	4.059	4.320	4.079
65	4.389	4.111	4.250	4.217
66	3.323	3.207	3.265	3.269
67	3.749	3.275	3.512	3.570
68	3.441	3.441	3.441	3.393
69	3.911	3.811	3.861	3.767
70	4.695	3.545	4.120	3.948
71	3.724	3.493	3.609	3.461
72	3.649	3.536	3.593	3.457
73	4.376	3.491	3.934	3.950
74	3.573	3.271	3.422	3.226
75	3.344	3.089	3.217	3.354
76	3.940	3.773	3.857	3.970
77	4.219	3.867	4.043	4.109
78	4.329	3.342	3.836	4.159
79	3.979	3.305	3.642	3.614
80	3.753	3.092	3.423	3.369
81	4.345	3.676	4.011	3.945
82	3.737	3.589	3.663	3.410
83	3.334	2.942	3.138	3.880
84	4.216	4.038	4.127	3.877
85	4.148	3.032	3.590	3.797
86	4.387	3.812	4.100	3.837
87	3.929	3.217	3.573	3.423
88	3.391	3.066	3.229	3.237
89	3.399	3.397	3.398	3.351
90	3.770	3.109	3.440	3.762
91	3.508	3.503	3.506	3.641
92	4.408	3.540	3.974	3.686
93	4.124	3.954	4.039	3.940
94	4.490	4.170	4.330	4.478
95	4.209	3.940	4.075	4.163
96	3.707	3.634	3.671	3.725
97	3.711	3.386	3.549	3.851
98	3.466	3.357	3.412	3.630
99	3.476	3.379	3.428	3.336
100	3.333	3.323	3.328	3.472
101	3.759	3.383	3.571	3.145
102	4.218	3.706	3.962	4.141

Table A.3 Characteristic strength of dewatered kaolin pellets from batch reactor at a rotation speed of 145 rpm

Pellet Nr.	d_{eq} (mm)	F_{max} (N)	τ_c (Nmm^{-2})
1	4.056	8.10	0.6266
2	4.535	5.02	0.3107
3	3.458	3.23	0.3438
4	3.509	4.79	0.4951
5	4.271	2.90	0.2023
6	4.019	5.22	0.4113
7	4.156	5.03	0.3706
8	3.755	6.29	0.5677
9	3.952	5.59	0.4555
10	3.937	7.00	0.5748
11	4.037	4.37	0.3413
12	3.658	4.00	0.3805
13	3.980	5.65	0.4515
14	3.839	5.77	0.4983
15	3.733	4.64	0.4238
16	3.874	5.70	0.4834
17	4.152	4.40	0.3248
18	3.857	4.06	0.3473
19	3.737	5.09	0.4639
20	4.018	4.36	0.3437
21	3.896	4.43	0.3714
22	3.873	8.00	0.6788
23	3.751	4.77	0.4315
24	3.823	5.79	0.5042
25	3.562	5.29	0.5306
26	3.615	4.50	0.4383
27	3.580	3.80	0.3774
28	3.844	4.40	0.3790
29	3.330	3.11	0.3569
30	3.800	4.80	0.4231
31	3.727	4.90	0.4490
32	3.695	4.08	0.3803
33	3.461	4.19	0.4452

Table A.4 Characteristic strength of dewatered kaolin pellets from batch reactor at a rotation speed of 165 rpm

Pellet Nr.	d_{eq} (mm)	F_{max} (N)	τ_c (Nmm^{-2})
1	3.066	4.40	0.5957
2	3.994	3.99	0.3183
3	4.297	5.22	0.3598
4	4.290	6.65	0.4599
5	3.838	4.57	0.3949
6	4.185	4.48	0.3255

(Continued)

Table A.4 (Continued).

Pellet Nr.	d_{eq} (mm)	F_{max} (N)	τ_c (Nmm^{-2})
7	3.560	4.09	0.4107
8	3.962	5.56	0.4508
9	4.220	5.78	0.4131
10	4.169	6.64	0.4862
11	3.923	4.80	0.3969
12	4.086	9.40	0.7166
13	4.016	5.00	0.3946
14	3.728	4.66	0.4267
15	4.044	4.39	0.3416
16	3.629	5.60	0.5412
17	3.997	5.23	0.4166
18	3.643	3.74	0.3587
19	3.918	4.98	0.4129
20	4.641	6.03	0.3563
21	4.037	5.50	0.4295
22	3.674	6.53	0.6157
23	4.184	6.15	0.4471
24	3.992	6.50	0.5191
25	4.203	4.97	0.3581
26	3.770	3.35	0.3000
27	3.382	3.40	0.3783
28	3.827	5.60	0.4866
29	3.777	4.69	0.4184
30	3.495	5.01	0.5220

References

Abu-Orf, M. M., & Dentel, S. K. (1997). Polymer Dose Assessment Using the Streaming Current Detector. *Water Environment Research*, 69(6), 1075–1085. http://doi.org/10.2175/106143097X125795

Abu-Orf, M., Muller, C. ., Park, C., & Novak, J. . (2004). Innovative Technologies to Reduce Water Content of Dewatered Municipal Residuals. *Journal of Residuals Science & Technology*, 1(2), 83–91.

Adachi, Y., Kobayashi, A., & Kobayashi, M. (2012). Structure of Colloidal Flocs in Relation to the Dynamic Properties of Unstable Suspension. *International Journal of Polymer Science*, 1–14. http://doi.org/10.1155/2012/574878

Adams, M. J., & McKeown, R. (1996). Micromechanical Analyses of the Pressure-Volume Relationship for Powders Under Confined uniaxial Compression. *Powder Technology*, 88(2), 155–163. http://doi.org/10.1016/0032-5910(96)03117-8

Addai-Mensah, J., & Prestidge, C. A. (2005). Structure Formation in Dispersed Systems. In B. Dobias & H. Stechemesser (Eds.), *Coagulation and Flocculation: Second Edition* (2nd ed., pp. 135–216). Boca Raton, FL: CRC Press.

Al Momani, F. A., & Örmeci, B. (2014). Optimization of Polymer Dose Based on Residual Polymer Concentration in Dewatering Supernatant. *Water, Air, & Soil Pollution*, 225(10), 1–11. http://doi.org/10.1007/s11270-014-2154-z

Amirtharajah, A., & Tambo, N. (1991). Mixing in Water Treatment. In A. Amirtharajah, M. M. Clark, & R. Trussell (Eds.), *Mixing in Coagulation and Flocculation* (pp. 3–34). Denver, CO: American Water Works Association.

André, B., Oshinowo, L. M., & Marshall, E. M. (2000). The Use of Large Eddy Simulation to Study Stirred Vessel Hydrodynamics (pp. 247–254). Presented at the 10th European Conference on Mixing, Delft, The Netherlands.

Argyropoulos, C. D., & Markatos, N. C. (2015). Recent Advances on the Numerical Modelling of Turbulent Flows. *Applied Mathematical Modelling*, 39(2), 693–732. http://doi.org/10.1016/j.apm.2014.07.001

Atkinson, J. F., Chakraborti, R. K., & Benschoten, J. E. (2005). Effects of Floc Size and Shape in Particle Aggregation. In S. N. Liss, I. G. Droppo, G. G. Leppard, & T. G. Milligan (Eds.), *Flocculation in Natural and Engineered Environmental Systems* (pp. 95–120). Boca Raton, FL: CRC Press.

Attia, Y. A. (1992). Flocculation. In J. S. Laskowski & J. Ralston (Eds.), *Colloid Chemistry in Mineral Processing* (pp. 277–308). Amsterdam: Elsevier.

Ayol, A., Dentel, S. K., & Filibeli, A. (2005). Dual Polymer Conditioning of Water Treatment Residuals. *Journal of Environmental Engineering*, 131(8), 1132–1138. http://doi.org/10.1061/(ASCE)0733-9372(2005)131:8(1132)

Ay, P., Hemme, A., Pflug, K., & Nitzsche, R. (1992). Application of Measurements of Charged State in Sludge Dewatering Technology. *Aufbereitungs-Technik*, 33(2), 57–66.

Bache, D. H. (2004). Floc Rupture and Turbulence: A Framework for Analysis. *Chemical Engineering Science*, 59(12), 2521–2534. http://doi.org/10.1016/j.ces.2004.01.055

Bache, D. H., & Gregory, R. (2007). *Flocs in Water Treatment*. London: IWA Publishing.

Bache, D. H., Papavasilopoulos, E. N., Rasool, E., Zhao, Y. Q., & McGilligan, J. F. (2003). Polymers in Alum Sludge Dewatering: Developments and Control. *Water and Environment Journal*, 17(2), 106–110. http://doi.org/10.1111/j.1747-6593.2003.tb00442.x

Bache, D. H., & Zhao, Y. Q. (2001). Optimising Polymer Use in Alum Sludge Conditioning: An Ad hoc Test. *Journal of Water Supply: Research and Technology - Aqua*, 50(1), 29–38.

Bagster, D. F. (1993). Aggregate Behaviour in Stirred Vessels. In A. P. Shamlou (Ed.), *Processing of Solid-Liquid Suspensions* (pp. 26–58). Oxford: Butterworth-Heinemann.

Bähr, S. (2006). *Experimental Studies of Fundamental Processes of Pelleting Flocculation* (PhD). Brandenburg University of Technology, Germany.

Bakker, A., & Oshinowo, L. M. (2004). Modelling of Turbulence in Stirred Vessels Using Large Eddy Simulation. *Chemical Engineering Research and Design*, 82(9), 1169–1178. http://doi.org/10.1205/cerd.82.9.1169.44153

Baldyga, J., & Bourne, J. R. (1984). A Fluid Mechanical Approach to Turbulent Mixing and Chemical Reaction Part II Micromixing in the Light of Turbulence Theory. *Chemical Engineering Communications*, 28(4-6), 243–258. http://doi.org/10.1080/00986448408940136

Bałdyga, J., & Bourne, J. R. (1999). *Turbulent Mixing and Chemical Reactions*. Weinheim: Wiley-VCH.

Bałdyga, J., & Pohorecki, R. (1995). Turbulent Micromixing in Chemical Reactors—A Review. *The Chemical Engineering Journal and the Biochemical Engineering Journal*, 58(2), 183–195. http://doi.org/10.1016/0923-0467(95)02982-6

Barbosa-Canovas, G., Ortega-Rivas, E., Juliano, P., & Yan, H. (2005). *Food Powders: Physical Properties, Processing, and Functionality*. New York: Kluwer Academic/Plenum Publishers.

Bartelt, A., Horn, D., Geiger, W., & Kern, G. (1994). Control and Optimization of Flocculation Processes in the Laboratory and in Plant. *Progress in Colloid & Polymer Science*, 95, 161–167. http://doi.org/10.1007/BFb0115718

Belfort, G. (1986). Fluid Mechanics and Cross-Flow Membrane Filtration. In H. S. Muralidhara (Ed.), *Advances in Solid-Liquid Separation* (pp. 165–189). Columbus, OH: Battelle Press.

Bemmer, G. G. (1979). *Agglomeration in Suspension: A Study of Mechanisms and Kinetics* (PhD). Delft University of Technology, Netherlands.

Benjamin, M. M., & Lawler, D. F. (2013). *Water Quality Engineering: Physical/Chemical Treatment Processes*. Hoboken, NJ: John Wiley & Sons.

Bergenstahl, B. (1995). Emulsions. In S. T. Beckett (Ed.), *Physico-Chemical Aspects of Food Processing* (pp. 49–64). Glasgow: Blackie Academic & Professional.

Besra, L., Sengupta, D. K., & Roy, S. K. (1998). Flocculant and Surfactant Aided Dewatering of Fine Particle Suspensions: A Review. *Mineral Processing and Extractive Metallurgy Review: An International Journal*, 18(1), 67–103. http://doi.org/10.1080/08827509808914153

Besra, L., Sengupta, D. K., Roy, S. K., & Ay, P. (2002). Polymer Adsorption: Its Correlation with Flocculation and Dewatering of Kaolin Suspension in the Presence and Absence of Surfactants. *International Journal of Mineral Processing*, 66(1-4), 183–202. http://doi.org/10.1016/S0301-7516(02)00064-9

Betatache, H., Aouabed, A., Drouiche, N., & Lounici, H. (2014). Conditioning of Sewage Sludge by Prickly Pear Cactus (Opuntia ficus Indica) Juice. *Ecological Engineering*, 70, 465–469. http://doi.org/10.1016/j.ecoleng.2014.06.031

Biggs, S. (2006). Aggregate Structures and Solid-Liquid Separation Processes. *KONA Powder and Particle Journal*, 24, 41–53.

Biswas, G., & Som, S. K. (2004). *Introduction to Fluid Mechanics and Fluid Machines*. New Delhi: Tata McGraw-Hill.

Böhm, N., & Kulicke, W.-M. (1997). Optimization of the Use of Polyelectrolytes for Dewatering Industrial Sludges of Various Origins. *Colloid and Polymer Science, 275*(1), 73–81. http://doi.org/10.1007/s003960050054

Boller, M., & Blaser, S. (1998). Particles Under Stress. *Water Science & Technology, 37*(10), 9–29.

Bolto, B. A. (2006). Coagulation and Flocculation with Organic Polyelectrolytes. In G. Newcombe & D. Dixon (Eds.), *Interface Science in Drinking Water Treatment: Theory and Applications* (Vol. 10, pp. 63–88). London: Academic Press.

Bouyer, D., Coufort, C., Liné, A., & Do-Quang, Z. (2005). Experimental Analysis of Floc Size Distributions in a 1-L Jar under Different Hydrodynamics and Physicochemical Conditions. *Journal of Colloid and Interface Science, 292*(2), 413–428. http://doi.org/10.1016/j.jcis.2005.06.011

Bouyer, D., Escudié, R., & Liné, A. (2005). Experimental Analysis of Hydrodynamics in a Jar-Test. *Process Safety and Environmental Protection, 83*(1), 22–30. http://doi.org/10.1205/psep.03109

Bouyer, D., Liné, A., & Do-Quang, Z. (2004). Experimental Analysis of Floc Size Distribution under Different Hydrodynamics in a Mixing Tank. *AIChE Journal, 50*(9), 2064–2081. http://doi.org/10.1002/aic.10242

Boyle, J. F., Manas-Zloczower, I., & Feke, D. L. (2005). Hydrodynamic Analysis of the Mechanisms of Agglomerate Dispersion. *Powder Technology, 153*(2), 127–133. http://doi.org/10.1016/j.powtec.2004.08.010

Bratby, J. (2006). *Coagulation and Flocculation in Water and Wastewater Treatment*. London: IWA Publishing.

Bridgeman, J., Jefferson, B., & Parsons, S. (2008). Assessing Floc Strength Using CFD to Improve Organics Removal. *Chemical Engineering Research and Design, 86*(8), 941–950. http://doi.org/10.1016/j.cherd.2008.02.007

Bridgeman, J., Jefferson, B., & Parsons, S. A. (2009). Computational Fluid Dynamics Modelling of Flocculation in Water Treatment: A Review. *Engineering Applications of Computational Fluid Mechanics, 3*(2), 220–241. http://doi.org/10.1080/19942060.2009.11015267

Bridgeman, J., Jefferson, B., & Parsons, S. A. (2010). The Development and Application of CFD Models for Water Treatment Flocculators. *Advances in Engineering Software, 41*(1), 99–109. http://doi.org/10.1016/j.advengsoft.2008.12.007

Brown, D. A. R., Jones, P. N., & Middleton, J. C. (2004). Experimental Methods. In E. L. Paul, V. A. Atiemo-Obeng, & S. M. Kresta (Eds.), *Handbook of Industrial Mixing: Science and Practice* (pp. 145–256). Hoboken, NJ: John Wiley & Sons.

Bubakova, P., Pivokonsky, M., & Filip, P. (2013). Effect of Shear Rate on Aggregate Size and Structure in the Process of Aggregation and at Steady State. *Powder Technology, 235*, 540–549. http://doi.org/10.1016/j.powtec.2012.11.014

Bugay, S., Escudié, R., & Liné, A. (2002). Experimental Analysis of Hydrodynamics in Axially Agitated Tank. *AIChE Journal, 48*(3), 463–475. http://doi.org/10.1002/aic.690480306

Byun, S., Kwon, J., Kim, M., Park, K., & Lee, S. (2007). Automatic Control of Polymer Dosage Using Streaming Potential for Waterworks Sludge Conditioning. *Separation and Purification Technology, 57*(2), 230–236. http://doi.org/10.1016/j.seppur.2007.03.009

Camp, T. R., & Stein, P. C. (1943). Velocity Gradients and Internal Work in Fluid Motion. *Journal of Boston Society of Civil Engineering, 30*, 219–237.

Carissimi, E., Miller, J. D., & Rubio, J. (2007). Characterization of the High Kinetic Energy Dissipation of the Flocs Generator Reactor (FGR). *International Journal of Mineral Processing, 85*(1-3), 41–49. http://doi.org/10.1016/j.minpro.2007.08.001

Carissimi, E., & Rubio, J. (2005). The Flocs Generator Reactor-FGR: A New Basis for Flocculation and Solid–Liquid Separation. *International Journal of Mineral Processing, 75*(3-4), 237–247. http://doi.org/10.1016/j.minpro.2004.08.021

Carissimi, E., & Rubio, J. (2015). Polymer-Bridging Flocculation Performance Using Turbulent Pipe Flow. *Minerals Engineering*, 70, 20–25. http://doi.org/10.1016/j.mineng.2014.08.019

Chakraborti, R. K., Atkinson, J. F., & Benschoten, J. E. (2000). Characterization of Alum Floc by Image Analysis. *Environmental Science & Technology*, 34(18), 3969–3976. http://doi.org/10.1021/es9908180

Colin, F., & Gazbar, S. (1995). Distribution of Water in Sludges in Relation to their Mechanical Dewatering. *Water Research*, 29(8), 2000–2005. http://doi.org/10.1016/0043-1354(94)00274-B

Concha, F. (2014). *Solid-Liquid Separation in the Mining Industry*. Heidelberg: Springer.

Coufort, C., Bouyer, D., & Liné, A. (2005). Flocculation Related to Local Hydrodynamics in a Taylor–Couette Reactor and in a Jar. *Chemical Engineering Science*, 60(8-9), 2179–2192. http://doi.org/10.1016/j.ces.2004.10.038

Craig, K., De Traversay, C., Bowen, B., Essemiani, K., Levecq, C., & Naylor, R. (2002). Hydraulic Study and Optimisation of Water Treatment Processes Using Numerical Simulation. *Water Science and Technology: Water Supply*, 2(5-6), 135–142.

Curran, S. J., & Black, R. A. (2005). Taylor-Vortex Bioreactors for Enhanced Mass Transport. In J. Chaudhuri & M. Al-Rubeai (Eds.), *Bioreactors for Tissue Engineering: Principles, Design and Operation* (pp. 47–85). Dordrecht: Springer.

Daniels, S. L. (1993). Flocculation. In J. McKetta (Ed.), *Unit Operations Handbook* (Vol. 2, pp. 140–174). New York, NY: Marcel Dekker.

Das, S., Bai, H., Wu, C., Kao, J.-H., Barney, B., Kidd, M., & Kuettel, M. (2016). Improving the Performance of Industrial Clarifiers Using Three-Dimensional Computational Fluid Dynamics. *Engineering Applications of Computational Fluid Mechanics*, 10(1), 130–144. http://doi.org/10.1080/19942060.2015.1121518

Dentel, S. K. (2010). Chemical Conditioning for Solid–Liquid Separation Processes. *Drying Technology: An International Journal*, 28(7), 843–849. http://doi.org/10.1080/07373937.2010.490490

Dentel, S. K., & Abu-Orf, M. M. (1995). Laboratory and Full-Scale Studies of Liquid Stream Viscosity and Streaming Current for Characterization and Monitoring of Dewaterability. *Water Research*, 29(12), 2663–2672. http://doi.org/10.1016/0043-1354(95)00142-8

Dentel, S. K., Abu-Orf, M. M., & Walker, C. A. (1998). Fundamental Methods for Optimizing Residuals Dewatering. In H. H. Hahn, E. Hoffmann, & H. Odegaard (Eds.), *Chemical Water and Wastewater Treatment V* (pp. 297–310). Heidelberg: Springer.

Dentel, S. K., Wehnes, K. M., & Abu-Orf, M. M. (1994). Use of Streaming Current and Other Parameters for Polymer Dose Control in Sludge Conditioning. In H. Hahn & R. Klute (Eds.), *Chemical Water and Wastewater Treatment III* (pp. 373–381). Heidelberg: Springer.

Dobias, B., & Von Rybinski, W. (1999). Stability of Dispersions. In B. Dobias, X. Qiu, & W. Von Rybinski (Eds.), *Solid-Liquid Dispersions* (pp. 244–278). New York, NY: Marcel Dekker.

Edwards, M. F., & Baker, M. R. (1997). A Review of Liquid Mixing Equipment. In N. Harnby, M. F. Edwards, & A. W. Nienow (Eds.), *Mixing in the Process Industries* (pp. 118–136). Oxford: Butterworth-Heinemann.

Edwards, M. F., Baker, M. R., & Godfrey, J. C. (1997). Mixing of Liquids in Stirred Tanks. In N. Harnby, M. F. Edwards, & A. W. Nienow (Eds.), *Mixing in the Process Industries* (pp. 137–158). Oxford: Butterworth-Heinemann.

Edzwald, J. A., Bottero, J. Y., Ives, K. J., & Klute, R. (1997). Particle Alteration and Particle Production Processes. In J. B. McEwen (Ed.), *Treatment Process Selection for Particle Removal* (pp. 73–122). Denver, CO: American Water Works Association.

Essemiani, K., & De Traversay, C. (2002). Optimisation of the Flocculation Process Using Computational Fluid Dynamics. In H. Hahn, E. Hoffman, & H. Odegaard (Eds.), *Chemical Water and Wastewater Treatment VII* (pp. 41–49). London: IWA Publishing.

Falk, L., & Commenge, J.-M. (2009). Characterization of Mixing and Segregation in Homogeneous Flow Systems. In V. Hessel, A. Renken, J. C. Schouten, & J.-I. Yoshida (Eds.), *Handbook of Micro Reactors* (pp. 147–171). Weinheim: John Wiley & Sons.

Farinato, R. S., Huang, S.-Y., & Hawkins, P. (1993). Polyelectrolyte-Assisted Dewatering. In R. S. Farinato & P. L. Dubin (Eds.), *Colloid-Polymer Interactions: From Fundamentals to Practice* (pp. 3–50). New York, NY: John Wiley & Sons.

Farrow, J. B., & Swift, J. D. (1996). A New Procedure for Assessing the Performance of Flocculants. *International Journal of Mineral Processing*, 46(1-3), 263–275. http://doi.org/10.1016/0301-7516(95)00084-4

Farrow, J., & Warren, L. (1993). Measurement of the Size of Aggregates in Suspension. In B. Dobias (Ed.), *Coagulation and Flocculation: Theory and Applications* (pp. 391–425). New York, NY: Marcel Dekker.

Gang, Z., Ting-lin, H., Chi, T., Li, Z., Wen-jie, H., Hong-da, H., & Chen, L. (2010). Settling Behaviour of Pellet Flocs in Pelleting Flocculation Process: Analysis through Operational Conditions. *Water Science & Technology*, 62(6), 1346–1352. http://doi.org/10.2166/wst.2010.429

Gillberg, L., Hanse, B., Karlsson, I., Enkel, A., & Palsson, A. (2003). *About Water Treatment*. (A. Lindquist, Ed.). Helsingborg: Kemira Kemwater.

Glasgow, L. (2005). Physicochemical Influences Upon Floc Deformability, Density, and Permeability (pp. 1–10). Presented at the 7th World Congress of Chemical Engineering, Glasgow, Scotland.

Godfrey, J. C., & Amirtharajah, A. (1991). Mixing in Liquids. In A. Amirtharajah, M. M. Clark, & R. Trussell (Eds.), *Mixing in Coagulation and Flocculation* (pp. 35–79). Denver, CO: American Water Works Association.

Goodwin, J. (2004). *Colloids and Interfaces with Surfactants and Polymers: An Introduction*. Hoboken, NJ: John Wiley & Sons.

Grasso, D., Subramaniam, K., Butkus, M., Strevett, K., & Bergendahl, J. (2002). A Review of Non-DLVO Interactions in Environmental Colloidal Systems. *Reviews in Environmental Science and Biotechnology*, 1(1), 17–38. http://doi.org/10.1023/A:1015146710500

Gregory, J. (1989). Fundamentals of Flocculation. *Critical Reviews in Environmental Control*, 19(3), 185–230. http://doi.org/10.1080/10643388909388365

Gregory, J. (1992). Flocculation of Fine Particles. In P. Mavros & K. A. Matis (Eds.), *Innovations in Flotation Technology* (pp. 101–124). Heidelberg: Springer Netherlands.

Gregory, J. (1993). Stability and Flocculation of Suspensions. In A. P. Shamlou (Ed.), *Processing of Solid-Liquid Suspensions* (pp. 59–92). Oxford: Butterworth-Heinemann.

Gregory, J. (1993). The Role of Colloid Interactions in Solid-Liquid Separation. *Water Science & Technology*, 27(10), 1–17.

Gregory, J. (2006a). Floc Formation and Floc Structure. In G. Newcombe & D. Dixon (Eds.), *Interface Science in Drinking Water Treatment: Theory and Applications* (Vol. 10, pp. 25–43). London: Academic Press.

Gregory, J. (2006b). *Particles in Water: Properties and Processes*. Boca Raton, FL: CRC Press.

Gregory, J. (2009). Monitoring Particle Aggregation Processes. *Advances in Colloid and Interface Science*, 147-148, 109–123. http://doi.org/10.1016/j.cis.2008.09.003

Gregory, J. (2013a). Flocculation Fundamentals. In T. Tadros (Ed.), *Encyclopedia of Colloid and Interface Science* (pp. 459–491). Heidelberg: Springer.

Gregory, J. (2013b). Flocculation Measurement Techniques. In T. Tadros (Ed.), *Encyclopedia of Colloid and Interface Science* (pp. 492–523). Heidelberg: Springer.

Gregory, J., & Guibai, L. (1991). Effects of Dosing and Mixing Conditions on Polymer Flocculation of Concentrated Suspensions. *Chemical Engineering Communications*, 108(1), 3–21. http://doi.org/10.1080/00986449108910948

Gregory, J., & Nelson, D. W. (1984). A New Optical Method for Flocculation Monitoring. In J. Gregory (Ed.), *Solid-Liquid Separation* (pp. 172–182). London: Ellis Horwood Ltd.

Guida, M., Mattei, M., Della Rocca, C., Melluso, G., & Meric, S. (2007). Optimization of Alum-Coagulation/Flocculation for COD and TSS Removal from Five Municipal Wastewater. *Desalination*, 211(1-3), 113–127. http://doi.org/10.1016/j.desal.2006.02.086

Hanson, A. T., & Cleasby, J. L. (1990). The Effects of Temperature on Turbulent Flocculation: Fluid Dynamics and Chemistry. *Journal American Water Works Association*, 82(11), 56–73.

Haralampides, K., McCorquodale, A. J., & Krishnappan, B. G. (2003). Deposition Properties of Fine Sediment. *Journal of Hydraulic Engineering*, 129(3), 230–234. http://doi.org/10.1061/(ASCE)0733-9429(2003)129:3(230)

Heath, A. R., & Koh, P. T. L. (2003). Combined Population Balance and CFD Modelling of Particle Aggregation by Polymeric Flocculant. Presented at the 3rd International Conference on CFD in the Minerals and Process Industries, Melbourne, Australia.

He, J., Liu, J., Yuan, Y., & Zhang, J. (2015). A Novel Quantitative Method for Evaluating Floc Strength Under Turbulent Flow Conditions. *Desalination and Water Treatment*, 56(7), 1975–1984. http://doi.org/10.1080/19443994.2014.958107

Hemme, A., Polte, R., & Ay, P. (1995). Pelleting Flocculation—The Alternative to Traditional Sludge Conditioning. *Aufbereitungs-Technik*, 36(5), 226–235.

Hendricks, D. W. (2011). *Fundamentals of Water Treatment Unit Processes: Physical, Chemical, and Biological*. Boca Raton, FL: CRC Press.

Herrington, T. M., Midmore, B. R., & Watts, J. C. (1993). Flocculation of Kaolin Suspensions by Polyelectrolytes. In P. L. Dubin & P. Tong (Eds.), *Colloid-Polymer Interactions: Particulate, Amphiphilic, and Biological Surfaces* (Vol. 532, pp. 161–181). San Francisco, CA: American Chemical Society.

Hessel, V., Hardt, S., & Löwe, H. (2004). *Chemical Micro Process Engineering: Fundamentals, Modelling and Reactions*. Weinheim: Wiley-VCH.

Hetherington, M., Pepperman, R., Shea, T. G., Stone, L., Reimers, R., & Rom, P. (2006). *Emerging Technologies for Biosolids Management* (No. EPA 832-R-06-005). Washington D.C.: U.S. Environmental Protection Agency.

Higashitani, K., & Kubota, T. (1987). Pelleting Flocculation of Colloidal Latex Particles. *Powder Technology*, 51(1), 61–69. http://doi.org/10.1016/0032-5910(87)80040-2

Higashitani, K., Shibata, T., & Matsuno, Y. (1987). Formation of Pellet Flocs from Kaolin Suspension and their Properties. *Journal of Chemical Engineering of Japan*, 20(2), 152–157.

Hjorth, M. (2009). *Flocculation and Solid-Liquid Separation of Animal Slurry: Fundamentals, Control and Application* (PhD). University of Southern Denmark, Denmark.

Hjorth, M., & Christensen, M. L. (2008). Evaluation of Methods to Determine Flocculation Procedure for Manure Separation. *Transactions of the ASABE*, 51(6), 2093–2103. http://doi.org/10.13031/2013.25391

Hjorth, M., Christensen, M. L., & Christensen, P. V. (2008). Flocculation, Coagulation, and Precipitation of Manure affecting Three Separation Techniques. *Bioresource Technology*, 99(18), 8598–8604. http://doi.org/10.1016/j.biortech.2008.04.009

Hjorth, M., & Jørgensen, B. U. (2012). Polymer Flocculation Mechanism in Animal Slurry Established by Charge Neutralization. *Water Research*, 46(4), 1045–1051. http://doi.org/10.1016/j.watres.2011.11.078

Hocking, M. B., Kimchuk, K. A., & Lowen, S. (1999). Polymeric Flocculants and Flocculation. *Journal of Macromolecular Science, Part C: Polymer Reviews*, 39(2), 177–203. http://doi.org/10.1081/MC-100101419

Hogg, R. (2000). Flocculation and Dewatering. *International Journal of Mineral Processing*, 58(1-4), 223–236. http://doi.org/10.1016/S0301-7516(99)00023-X

Hogg, R. (2005). Flocculation and Dewatering of Fine-Particle Suspension. In B. Dobias & H. Stechemesser (Eds.), *Coagulation and Flocculation: Second Edition* (2nd ed., pp. 805–850). Boca Raton, FL: CRC Press.

Ives, K. J. (1984). Experiments in Orthokinetic Flocculation. In J. Gregory (Ed.), *Solid-Liquid Separation* (pp. 196–220). London: Ellis Horwood Ltd.

Jarvis, P., Jefferson, B., Gregory, J., & Parsons, S. A. (2005). A Review of Floc Strength and Breakage. *Water Research, 39*(14), 3121–3137. http://doi.org/10.1016/j.watres.2005.05.022

Kharoua, N., Khezzar, L., & Saadawi, H. (2013). CFD Modelling of a Horizontal Three-Phase Separator: A Population Balance Approach. *American Journal of Fluid Dynamics, 3*(4), 2168–4715. http://doi.org/10.5923/j.ajfd.20130304.03

Kim, S., & Palomino, A. M. (2009). Polyacrylamide-Treated Kaolin: A Fabric Study. *Applied Clay Science, 45*(4), 270–279. http://doi.org/10.1016/j.clay.2009.06.009

Kissa, E. (1999). *Dispersions: Characterization, Testing, and Measurement*. New York, NY: Marcel Dekker.

Kobayashi, M., Adachi, Y., & Ooi, S. (1999). Breakup of Fractal Flocs in a Turbulent Flow. *Langmuir, 15*(13), 4351–4356. http://doi.org/10.1021/la9807630

Kockmann, N. (2008). *Transport Phenomena in Micro Process Engineering*. Heidelberg: Springer.

Kolmogorov, A. N. (1991a). Dissipation of Energy in the Locally Isotropic Turbulence. *Proceedings A, 434*(1890), 15–17. http://doi.org/10.1098/rspa.1991.0076

Kolmogorov, A. N. (1991b). The Local Structure of Turbulence in Incompressible Viscous Fluid for Very Large Reynolds Numbers. *Proceedings A, 434*(1890), 9–13. http://doi.org/10.1098/rspa.1991.0075

Korpijärvi, J., Laine, E., & Ahlstedt, H. (2000). Using CFD in the Study of Mixing in Coagulation and Flocculation. In H. H. Hahn, E. Hoffmann, & H. Odegaard (Eds.), *Chemical Water and Wastewater Treatment VI* (pp. 89–99). Heidelberg: Springer.

Kramer, T. A., & Clark, M. M. (1997). Influence of Strain-Rate on Coagulation Kinetics. *Journal of Environmental Engineering, 123*(5), 444–452. http://doi.org/10.1061/(ASCE)0733-9372(1997)123:5(444)

Kramer, T. A., & Clark, M. M. (1999). Incorporation of Aggregate Breakup in the Simulation of Orthokinetic Coagulation. *Journal of Colloid and Interface Science, 216*(1), 116–126. http://doi.org/10.1006/jcis.1999.6305

Kresta, S. M., & Brodkey, R. S. (2004). Turbulence in Mixing Applications. In E. L. Paul, V. A. Atiemo-Obeng, & S. M. Kresta (Eds.), *Handbook of Industrial Mixing: Science and Practice* (pp. 19–87). Hoboken, NJ: John Wiley & Sons.

Kruster, K. A. (1991). *The Influence of Turbulence on Aggregation of Small Particles in Agitated Vessels* (PhD). Technische Universiteit Eindhoven, Netherlands.

Lagaly, G. (1993). From Clay Mineral Crystals to Colloidal Clay Mineral Dispersions. In B. Dobias (Ed.), *Coagulation and Flocculation: Theory and Applications* (pp. 427–494). New York, NY: Marcel Dekker.

Lagaly, G. (2005). From Clay Mineral Crystals to Colloidal Clay Mineral Dispersions. In B. Dobias & H. Stechemesser (Eds.), *Coagulation and Flocculation: Second Edition* (2nd ed., pp. 519–600). Boca Raton, FL: CRC Press.

Laskowski, J. S., & Pugh, R. J. (1992). Dispersions Stability and Dispersing Agents. In J. S. Laskowski & J. Ralston (Eds.), *Colloid Chemistry in Mineral Processing* (pp. 115–170). Amsterdam: Elsevier.

Lawler, F. D. (1993). Physical Aspects of Flocculation: From Microscale to Macroscale. *Water Research, 27*(10), 165–180.

Lebovka, N. I. (2013). Aggregation of Charged Colloidal Particles. In M. Müller (Ed.), *Polyelectrolyte Complexes in the Dispersed and Solid State I* (pp. 57–96). Heidelberg: Springer.

Lee, B. J., & Molz, F. (2014). Numerical Simulation of Turbulence-Induced Flocculation and Sedimentation in a Flocculant-Aided Sediment Retention Pond. *Environmental Engineering Research*, 19(2), 165–174. http://doi.org/10.4491/eer.2014.19.2.165

Lee, C. H., & Liu, J. C. (2000). Enhanced Sludge Dewatering by Dual Polyelectrolytes Conditioning. *Water Research*, 34(18), 4430–436. http://doi.org/10.1016/S0043-1354(00)00209-8

Lee, D. ., & Hsu, Y. . (1995). Measurement of Bound Water in Sludges: A Comparative Study. *Water Environment Research*, 67(3), 310–317.

Lee, K. E., Morad, N., Teng, T. T., & Poh, B. T. (2012). Development, Characterization and the Application of Hybrid Materials in Coagulation/Flocculation of Wastewater: A Review. *Chemical Engineering Journal*, 203(1), 370–386. http://doi.org/10.1016/j.cej.2012.06.109

Letterman, R. D., Amirtharajah, A., & O'Meila, C. R. (2010). Coagulation and Flocculation. In J. Edzwald (Ed.), *Water Quality & Treatment: A Handbook on Drinking Water* (pp. 6.1–6.66). New York, NY: McGraw- Hill.

Lick, W. (2008). *Sediment and Contaminant Transport in Surface Waters*. Boca Raton, FL: CRC Press.

Lick, W., Huang, H., & Jepsen, R. (1993). Flocculation of Fine-Grained Sediments Due to Differential Settling. *Journal of Geophysical Research Oceans*, 98(6), 10279–10288. http://doi.org/10.1029/93JC00519

Lick, W., & Lick, J. (1988). Aggregation and Disaggregation of Fine-Grained Lake Sediments. *Journal of Great Lakes Research*, 14(4), 514–523. http://doi.org/10.1016/S0380-1330(88)71583-X

Lick, W., Lick, J., & Ziegler, C. K. (1992a). Flocculation and its Effect of the Vertical Transport of Fine-Grained Sediments. In B. T. Hart & P. G. Sly (Eds.), *Sediment/Water Interactions* (pp. 1–16). Heidelberg: Springer.

Lick, W., Lick, J., & Ziegler, C. K. (1992b). Flocculation and its Effect of the Vertical Transport of Fine-Grained Sediments. *Hydrobiologia*, 235-236(1), 1–16. http://doi.org/10.1007/BF00026196

Liu, S. X., & Glasgow, L. A. (1997). Aggregate Disintegration in Turbulent Jets. *Water, Air, and Soil Pollution*, 95(1-4), 257–275. http://doi.org/10.1007/BF02406169

Logan, B. E. (2012). *Environmental Transport Processes*. Hoboken, NJ: John Wiley & Sons.

Lu, S., Ding, Y., & Guo, J. (1998). Kinetics of Fine Particle Aggregation in Turbulence. *Advances in Colloid and Interface Science*, 78(3), 197–235. http://doi.org/10.1016/S0001-8686(98)00062-1

Maggi, F. (2005). *Flocculation Dynamics of Cohesive Sediment* (PhD). Delft Universty of Technology, Netherlands.

Mahmoud, A., Olivier, J., Vaxelaire, J., & Hoadley, A. F. (2010). Electrical Field: A Historical Review of its Application and Contributions in Wastewater Sludge Dewatering. *Water Research*, 44(8), 2381–2407. http://doi.org/10.1016/j.watres.2010.01.033

Mahmoud, A., Olivier, J., Vaxelaire, J., & Hoadley, A. F. A. (2013). Advances in Mechanical Dewatering of Wastewater Sludge Treatment. In S. K. Sharma & R. Sanghi (Eds.), *Wastewater Reuse and Management* (pp. 253–303). Heidelberg: Springer.

Marchisio, D. L., Vigil, D. R., & Fox, R. O. (2003). Implementation of the Quadrature Method of Moments in CFD Codes for Aggregation–Breakage Problems. *Chemical Engineering Science*, 58(15), 3337–3351. http://doi.org/10.1016/S0009-2509(03)00211-2

Marshall, E. M., & Bakker, A. (2004). Computational Fluid Mixing. In E. L. Paul, V. A. Atiemo-Obeng, & S. M. Kresta (Eds.), *Handbook of Industrial Mixing: Science and Practice* (pp. 257–343). Hoboken, NJ: John Wiley & Sons.

Marshall, J. S., & Li, S. (2014). *Adhesive Particle Flow: A Discrete-Element Approach*. New York, NY: Cambridge University Press.

Mavros, P. (2001). Flow Visualization in Stirred Vessels: A Review of Experimental Techniques. *Chemical Engineering Research and Design*, 79(2), 113–127. http://doi.org/10.1205/02638760151095926

McConnachie, G. (1991). Turbulence Intensity of Mixing in Relation to Flocculation. *Journal of Environmental Engineering*, *117*(6), 731–750. http://doi.org/10.1061/(ASCE)0733-9372(1991)117:6(731)

Mhaisalkar, V. A., Paramasivam, R., & Bhole, A. G. (1986). An Innovative Technique for Determining Velocity Gradient in Coagulation-Flocculation Process. *Water Research*, *20*(10), 1307–1314. http://doi.org/10.1016/0043-1354(86)90162-4

Mikkelsen, L. H. (2003). Applications and Limitations of the Colloid Titration Method for Measuring Activated Sludge Surface Charges. *Water Research*, *37*(10), 2458–2466. http://doi.org/10.1016/S0043-1354(03)00021-6

Milligan, T. G., & Hill, P. S. (1998). A Laboratory Assessment of the Relative Importance of Turbulence, Particle Composition, and Concentration in Limiting Maximal Floc Size and Settling Behaviour. *Journal of Sea Research*, *39*(3-4), 227–241. http://doi.org/10.1016/S1385-1101(97)00062-2

Moody, G., & Norman, P. (2005). Chemical Pre-treatment. In S. Tarleton & R. Wakeman (Eds.), *Solid-Liquid Separation: Scale-Up of Industrial Equipment* (pp. 38–81). Oxford: Elsevier.

Moudgil, B. M. (1986). Selection of Flocculants for Solid-Liquid Separation Process. In H. S. Muralidhara (Ed.), *Advances in Solid-Liquid Separation* (pp. 191–204). Columbus, OH: Battelle Press.

Mowla, D., Tran, H. N., & Allen, D. G. (2013). A Review of the Properties of Biosludge and its Relevance to Enhanced Dewatering Processes. *Biomass and Bioenergy*, *58*, 365–378. http://doi.org/10.1016/j.biombioe.2013.09.002

Mühle, K. (1993). Floc Stability in Laminar and Turbulent Flow. In B. Dobias (Ed.), *Coagulation and Flocculation: Theory and Applications* (pp. 355–390). New York, NY: Marcel Dekker.

Myers, K. J., Ward, R. W., & Baker, A. (2014). A Digital Particle Image Velocimetry Investigation of Flow Field Instabilities of Axial-Flow Impellers. *Journal of Fluids Engineering*, *119*(3), 623–632. http://doi.org/10.1115/1.2819290

Neumann, L. E., & Howes, T. (2007). Aggregation and Breakage Rates in the Flocculation of Estuarine Cohesive Sediments. In J. P. Y. Maa, L. P. Sanford, & D. H. Schoellhamer (Eds.), *Estuarine and Coastal Fine Sediment Dynamics* (pp. 35–53). Amsterdam: Elsevier.

Nopens, I. (2005). *Modelling the Activated Sludge Flocculation Process: A Population Balance Approach* (PhD). University of Ghent, Belgium.

Nopens, I. (2007). Improved Prediction of Effluent Suspended Solids in Clarifiers through Integration of a Population Balance Model. Presented at the IWA Particle Separation Conference, Toulouse, France.

Novak, J. T. (2001). Dewatering. In L. Spinosa & A. Vesilind (Eds.), *Sludge Into Biosolids: Processing, Disposal and Utilization* (pp. 339–363). London: IWA Publishing.

Novak, T. . (2006). Dewatering of Sewage Sludge. *Drying Technology*, *24*(10), 1257–1262. http://doi.org/10.1080/07373930600840419

Oldshue, J. Y., & Trussell, R. R. (1991). Design of Impellers for Mixing. In A. Amirtharajah, M. M. Clark, & R. Trussell (Eds.), *Mixing in Coagulation and Flocculation* (pp. 309–342). Denver, CO: American Water Works Association.

Oshinowo, L., Elsaadawy, E., & Vilagines, R. (2015). CFD Modeling of Oil-Water Separation Efficiency in Three-Phase Separators. In J. E. Olsen & S. T. Johansen (Eds.), *Progress in Applied CFD* (pp. 207–216). Oslo: SINTEF Academic Press.

Oshinowo, L. M., Quintero, C. G., & Vilagines, R. D. (2016). CFD and Population Balance Modeling of Crude Oil Emulsions in Batch Gravity Separation – Comparison to Ultrasound Experiments. *Journal of Dispersion Science and Technology*, *37*(5), 665–675. http://doi.org/10.1080/01932691.2015.1054508

Outwater, A., & Tansel, B. (1994). *Reuse of Sludge and Minor Wastewater Residuals*. Boca Raton, FL: CRC Press.

Oyegbile, B., Ay, P., & Narra, S. (2016). Optimisation of Micro- Processes for Shear Assisted Solid-Liquid Separation in a Rotatory Batch Flow Vortex Reactor. *Journal of Water Reuse and Desalination*, 6(1). http://doi.org/10.2166/wrd.2015.057

Oyegbile, B., Ay, P., & Narra, S. (2016). Optimization of Physicochemical Process for Pretreatment of Fine Suspension by Flocculation Prior to Dewatering. *Desalination and Water Treatment*, 57(6), 2726–2736. http://doi.org/10.1080/19443994.2015.1043591

Panswad, T., & Polwanich, S. (1998). Pilot Plant Application of Pelletisation Process on Low-Turbidity River Water. *Journal of Water Supply: Research and Technology-AQUA*, 47(5), 236–244.

Papavasilopoulos, E. N. (1997). Viscosity as a Criterion for Optimum Dosing of Polymers in Waterworks Sludges. *Water and Environment Journal*, 11(3), 217–224. http://doi.org/10.1111/j.1747-6593.1997.tb00118.x

Partheniades, E. (1993). Turbulence, Flocculation and Cohesive Sediment Dynamics. In A. J. Mehta (Ed.), *Nearshore and Estuarine Cohesive Sediment Transport* (pp. 40–59). Washington DC: American Geophysical Union.

Partheniades, E. (2009). *Cohesive Sediments in Open Channels: Properties, Transport and Applications*. Oxford: Butterworth-Heinemann.

Pechlivanidis, G., Keramaris, E., & Pechlivanidis, I. (2014). Experimental Study of the Effects of Grass Vegetation and Gravel Bed on the Turbulent Flow Using Particle Image Velocimetry. *Journal of Turbulence*, 16(1), 1–14. http://doi.org/10.1080/14685248.2014.946605

Peng, S. J., & Williams, R. A. (1993). Control and Optimisation of Mineral Flocculation and Transport Processes Using On-line Particle Size Analysis. *Minerals Engineering*, 6(2), 133–153. http://doi.org/10.1016/0892-6875(93)90128-A

Perrard, M., Le Sauze, N., Xuereb, C., & Bertrand, J. (2000). Characterisation of the Turbulence in a Stirred Tank Using Particle Image Velocimetry. In H. E. A. Van den Akker & J. J. Derksen (Eds.), *10th European Conference on Mixing* (pp. 345–352). Amsterdam: Elsevier.

Petzold, G. (1993). Dual-Addition Schemes. In R. S. Farinato & P. L. Dubin (Eds.), *Colloid-Polymer Interactions: From Fundamentals to Practice* (pp. 83–100). New York, NY: John Wiley & Sons.

Petzold, G., & Schwarz, S. (2013). Polyelectrolyte Complexes in Flocculation Applications. In M. Müller (Ed.), *Polyelectrolyte Complexes in the Dispersed and Solid State II* (pp. 25–65). Heidelberg: Springer.

Pietsch, W. (2002). *Agglomeration Processes: Phenomena, Technologies, Equipment*. Weinheim: Wiley-VCH.

Popa, I., Papastavrou, G., & Borkovec, M. (2010). Charge Regulation Effects on Electrostatic Patch-Charge Attraction Induced by Adsorbed Dendrimers. *Physical Chemsitry Chemical Physics*, 12(18), 4863–4871. http://doi.org/10.1039/B925812D

Prat, O. O., & Ducoste, J. J. (2007). Simulation of Flocculation in Stirred Vessels Lagrangian Versus Eulerian. *Chemical Engineering Research and Design*, 85(2), 207–219. http://doi.org/10.1205/cherd05001

Prat, O. P., & Ducoste, J. J. (2006). Modeling Spatial Distribution of Floc Size in Turbulent Processes Using the Quadrature Method of Moment and Computational Fluid Dynamics. *Chemical Engineering Science*, 61(1), 75–86. http://doi.org/10.1016/j.ces.2004.11.070

Raffel, M., Willert, C. E., Wereley, S. T., & Kompenhans, J. (2007). *Particle Image Velocimetry: A Practical Guide*. Heidelberg: Springer.

Ramphal, S., & Sibiya, S. M. (2014). Optimization of Time Requirement for Rapid Mixing During Coagulation Using a Photometric Dispersion Analyzer. *Procedia Engineering*, 70, 1401–1410. http://doi.org/10.1016/j.proeng.2014.02.155

Richter, H. J., Josbt, K., Friedrich, K., Heinze, W., Friedrich, E., & Hermel, W. (1995). Possible Applications of Streaming Potential Measurements in Relation to Dewatering of Waste Water Sludge. *Aufbereitungs-Technik*, 34(5), 257–262.

Rulyov, N. N. (2010). Physicochemical Microhydrodynamics of Ultradisperse Systems. In V. M. Starov (Ed.), *Nanoscience: Colloidal and Interfacial Aspects* (pp. 969–995). Boca Raton, FL: CRC Press.

Runkana, V., Somasundaran, P., & Kapur, P. C. (2006). A Population Balance Model for Flocculation of Colloidal Suspensions by Polymer Bridging. *Chemical Engineering Science, 61*(1), 182–191. http://doi.org/10.1016/j.ces.2005.01.046

Sabah, E., & Erkan, Z. E. (2006). Interaction Mechanism of Flocculants with Coal Waste Slurry. *Fuel, 85*(3), 350–359. http://doi.org/10.1016/j.fuel.2005.06.005

Samaras, K., Zouboulis, A., Karapantsios, T., & Kostoglou, M. (2010). A CFD-Based Simulation Study of a Large Scale Flocculation Tank for Potable Water Treatment. *Chemical Engineering Journal, 162*(1), 208–216. http://doi.org/10.1016/j.cej.2010.05.032

Sanin, F. D., Clarkson, W. W., & Vesilind, P. A. (2011). *Sludge Engineering: The Treatment and Disposal of Wastewater Sludges*. Lancaster, PA: DEStech Publications.

Schramm, L. L. (2005). *Emulsions, Foams, and Suspensions*. Weinheim: Wiley VCH.

Serra, T., & Casamitjana, X. (1997). Modelling the Aggregation and Break-up of Fractal Aggregates in a Shear Flow. *Applied Scientific Research, 59*(2-3), 255–268. http://doi.org/10.1023/A:1001143707607

Shamlou, A. P., & Hooker-Titchener, N. (1993). Turbulent Aggregation and Breakup of Particles in Liquids in Stirred Vessels. In A. P. Shamlou (Ed.), *Processing of Solid-Liquid Suspensions* (pp. 1–25). Oxford: Butterworth-Heinemann.

Shammas, N. K. (2005). Coagulation and Flocculation. In L. K. Wang, Y.-T. Hung, & N. K. Shammas (Eds.), *Physicochemical Treatment Processes* (Vol. 3, pp. 103–139). Totowa, NJ: Humana Press.

Sher, F., Malik, A., & Liu, H. (2013). Industrial Polymer Effluent Treatment by Chemical Coagulation and Flocculation. *Journal of Environmental Chemical Engineering, 1*(4), 684–689. http://doi.org/10.1016/j.jece.2013.07.003

Sievers, M., Stoll, S. M., Schroeder, C., Niedermeiser, M., & Onyeche, T. I. (2008). Sludge Dewatering and Aggregate Formation Effects through Taylor Vortex Assisted Flocculation. *Separation Science and Technology, 43*(7), 1595–1609. http://doi.org/10.1080/01496390801973888

Singh, S. (2012). *Experiments in Fluid Mechanics*. New Delhi: PHI Learning.

Smith-Palmer, T., & Pelton, R. (2006). Flocculation of Particles. In P. Somasundaran (Ed.), *Encyclopedia of Surface and Colloid Science* (5th ed., pp. 2584–2599). Boca Raton, FL: CRC Press.

Son, M., & Hsu, T.-J. (2008). Flocculation Model of Cohesive Sediment Using Variable Fractal Dimension. *Environmental Fluid Mechanics, 8*(1), 55–71. http://doi.org/10.1007/s10652-007-9050-7

Soos, M., Moussa, A. S., Ehrl, L., Sefcik, J., Wu, H., & Morbidelli, M. (2008). Effect of Shear Rate on Aggregate Size and Morphology Investigated Under Turbulent Conditions in Stirred Tank. *Journal of Colloid and Interface Science, 319*(2), 577–589. http://doi.org/10.1016/j.jcis.2007.12.005

Sparks, T. (1996). *Fluid Mixing in Rotor/Stator Mixers* (PhD). Cranfield University, UK.

Spicer, P. T. (1997). *Shear-Induced Aggregation-Fragmentation: Mixing and Aggregate Morphology Effects* (PhD). University of Cincinnati, USA.

Spicer, P. T., & Pratsinis, S. E. (1996). Shear-Induced Flocculation: The Evolution of Floc Structure and the Shape of the Size Distribution at Steady State. *Water Research, 30*(5), 1049–1056. http://doi.org/10.1016/0043-1354(95)00253-7

Svarovsky, L. (2000). *Solid-Liquid Separation* (4th ed.). Woburn, MA: Butterworth-Heinemann.

Taboada-Serrano, P., Chin, C.-J., Yiacoumi, S., & Tsouris, C. (2005). Modeling Aggregation of Colloidal Particles. *Current Opinion in Colloid & Interface Science, 10*(3-4), 123–132. http://doi.org/10.1016/j.cocis.2005.07.003

Tambo, N. (1990). Optimization of Flocculation in Connection with Various Solid-Liquid Separation Processes. In H. Hahn & R. Klute (Eds.), *Chemical Water and Wastewater Treatment* (pp. 17–32). Heidelberg: Springer.

Tambo, N., & François, R. J. (1991). Mixing, Breakup and Floc Characteristics. In A. Amirtharajah, M. M. Clark, & R. Trussell (Eds.), *Mixing in Coagulation and Flocculation* (pp. 256–281). Denver, CO: American Water Works Association.

Tambo, N., & Wang, C. C. (1993). The Mechanism of Pellet Flocculation in Fluidized-Bed Operations. *Journal of Water Supply: Research and Technology-AQUA*, 42(2), 67–76.

Tarleton, S., & Wakeman, R. (2006). *Solid/Liquid Separation: Equipment Selection and Process Design*. Oxford: Elsevier.

Tchobanoglous, G., Burton, F. L., & Stensel, D. H. (2003). *Wastewater Engineering: Treatment and Reuse*. New York, NY: Mc-Graw Hill.

Teefy, S., Farmerie, J., & Pyles, E. (2001). Jar Test. In American Water Works Association (Ed.), *Operational Control of Coagulation and Filtration Processes* (3rd ed., pp. 17–57). Denver, CO: American Water Works Association.

Thoenes, D. (1998). *Chemical Reactor Development: From Laboratory Synthesis to Industrial Production*. Dordrecht: Springer.

Thomas, D. N., Judd, S. J., & Fawcett, N. (1999). Flocculation Modelling: A Review. *Water Research*, 33(7), 1579–1592. http://doi.org/10.1016/S0043-1354(98)00392-3

Thomas, S. F., Rooks, P., Rudin, F., Cagney, N., Balabani, S., Atkinson, S., … Allen, M. J. (2015). Swirl Flow Bioreactor Containing Dendritic Copper-Containing Alginate Beads: A Potential Rapid Method for the Eradication of Escherichia Coli from Waste Water Streams. *Journal of Water Process Engineering*, 5, 6–14. http://doi.org/10.1016/j.jwpe.2014.10.010

Tooby, P. F., Wick, G. L., & Isaacs, J. D. (1977). The Motion of a Small Sphere in a Rotating Velocity Field: A Possible Mechanism for Suspending Particles in Turbulence. *Journal of Geophysical Research*, 82(15), 2096–2100. http://doi.org/10.1029/JC082i015p02096

Torfs, E. (2015). *Different Settling Regimes in Secondary Settling Tanks: Experimental Process Analysis, Model Development and Calibration* (PhD). Ghent University, Belgium.

Torfs, E., Vesvikar, M., & Nopens, I. (2013). Improved Predictions of Effluent Suspended Solids in Wastewater Treatment Plants by Integration of a PBM with Computational Fluid Dynamics. Presented at the 5th Population Balance Modelling Conference, Bangalore, India.

Tsai, C.-H., Iacobellis, S., & Lick, W. (1987). Flocculation of Fine-Grained Lake Sediments Due to a Uniform Shear Stress. *Journal of Great Lakes Research*, 13(2), 135–146. http://doi.org/10.1016/S0380-1330(87)71637-2

Utomo, A., Baker, M., & Pacek, A. W. (2009). The Effect of Stator Geometry on the Flow Pattern and Energy Dissipation Rate in a Rotor-Stator Mixer. *Chemical Engineering Research and Design*, 87(4), 533–542. http://doi.org/10.1016/j.cherd.2008.12.011

Utomo, A. T., Baker, M., & Pacek, A. W. (2008). Flow Pattern, Periodicity and Energy Dissipation in a Batch Rotor–Stator Mixer. *Chemical Engineering Research and Design*, 86(12), 1397–1409. http://doi.org/10.1016/j.cherd.2008.07.012

Van Leussen, W. (2011). Aggregation of Particles, Settling Velocity of Mud Flocs-A Review. In J. Dronkers & W. Van Leussen (Eds.), *Physical Processes in Estuarine* (pp. 347–403). Heidelberg: Springer.

Vesilind, A. (1988). Capillary Suction Time as a Fundamental Measure of Sludge Dewaterability. *Journal Water Pollution Control Federation*, 60(2), 215–220.

Vesilind, P. A. (1994). The Role of Water in Sludge Dewatering. *Water Environment Research*, 66(1), 4–11.

Vesilind, P. ., & Tsang, K. . (1990). Moisture Distribution in Sludges. *Water Science & Technology*, 22(12), 135–142.

Vigdergauz, V. E., & Gol'berg, G. Y. (2012). Kinetics of Mechanical Floccule Synaeresis. *Journal of Mining Science*, 48(2), 347–353. http://doi.org/10.1134/S1062739148020165

Visscher, F., Van der Schaaf, J., Nijhuis, T. A., & Schouten, J. C. (2013). Rotating Reactors – A Review. *Chemical Engineering Research and Design*, 91(10), 1923–1940. http://doi.org/10.1016/j.cherd.2013.07.021

Von Homeyer, A., Krentz, D. O., Kulicke, W. M., & Lerche, D. (1999). Optimization of the Polyelectrolyte Dosage for Dewatering Sewage Sludge Suspensions by Means of a New Centrifugation Analyser with an Optoelectronic Sensor. *Colloid & Polymer Science*, 277(7), 637–645. http://doi.org/10.1007/s003960050435

Wakeman, R. J. (2007). Separation Technologies for Sludge Dewatering. *Journal of Hazardous Materials*, 144(3), 614–619. http://doi.org/10.1016/j.jhazmat.2007.01.084

Walaszek, W. (2007). *Investigation Upon Structure of Pellet Flocs against Process Performance as a Tool to Optimize Sludge Conditioning* (PhD). Brandenburg University of Technology, Germany.

Walaszek, W., & Ay, P. (2005). Pelleting Flocculation—An Alternative Technique to Optimise Sludge Conditioning. *International Journal of Mineral Processing*, 76(3), 173–180. http://doi.org/10.1016/j.minpro.2005.01.001

Walaszek, W., & Ay, P. (2006a). Extended Interpretation of the Structural Attributes of Pellet Flocs in Pelleting Flocculation. *Minerals Engineering*, 19(13), 1397–1400. http://doi.org/10.1016/j.mineng.2006.03.004

Walaszek, W., & Ay, P. (2006b). Porosity and Interior Structure Analysis of Pellet-Flocs. *Colloids and Surfaces A: Physicochemical and Engineering Aspects*, 280(1-3), 155–162. http://doi.org/10.1016/j.colsurfa.2006.01.049

Wang, G., Zhou, S., Joshi, J. B., Jameson, G. J., & Evans, G. M. (2014). An Energy Model on Particle Detachment in the Turbulent Field. *Minerals Engineering*, 69, 165–169. http://doi.org/10.1016/j.mineng.2014.07.018

Wang, L., Marchisio, D. L., Vigil, R. D., & Fox, R. O. (2005). CFD Simulation of Aggregation and Breakage Processes in Laminar Taylor–Couette Flow. *Journal of Colloid and Interface Science*, 282(2), 380–396. http://doi.org/10.1016/j.jcis.2004.08.127

Wang, X. H., & Jiang, C. (2006). Papermaking Part II: Surface and Colloid Chemsitry of Papermaking Process. In P. Somasundaran (Ed.), *Encyclopedia of Surface and Colloid Science* (5th ed., pp. 4435–4451). Boca Raton, FL: CRC Press.

Wang, X. ., Jin, P. ., Yuan, H. ., Wang, E. ., & Tambo, N. (2004). Pilot Study of a Fluidized-Pellet-Bed Technique for Simultaneous Solid/liquid Separation and Sludge Thickening in a Sewage Treatment Plant. *Water Science & Technology*, 49(1), 81–88.

Ward-Smith, J., & Massey, B. (2006). *Mechanics of Fluid* (8th ed.). New York: CRC Press.

Watanabe, Y., & Tanaka, K. (1999). Innovative Sludge Handling through Pelletization/Thickening. *Water Research*, 33(15), 3245–3252. http://doi.org/10.1016/S0043-1354(99)00045-7

Wilkinson, K. J., & Reinhardt, A. (2005). Contrasting Roles of Natural Organic Matter on Colloidal Stabilization and Flocculation. In S. N. Liss, I. G. Droppo, G. G. Leppard, & T. G. Milligan (Eds.), *Flocculation in Natural and Engineered Environmental Systems* (pp. 143–170). Boca Raton, FL: CRC Press.

Wimmer, M. (1995). An Experimental Investigation of Taylor Vortex Flow between Conical Cylinders. *Journal of Fluid Mechanics*, 292, 205– 227. http://doi.org/10.1017/S0022112095001492

Wimmer, M., & Zierep, J. (2000). Transition from Taylor Vortices to Cross-flow Instabilities. *Acta Mechanica*, 140(1), 17–30. http://doi.org/10.1007/BF01175977

Winterwerp, J. C. (1998). A Simple Model for Turbulence Induced Flocculation of Cohesive Sediment. *Journal of Hydraulic Research*, 36(3), 309–326. http://doi.org/10.1080/00221689809498621

Wu, C. C., Huang, C., & Lee, D. J. (1998). Bound Water Content and Water Binding Strength on Sludge Flocs. *Water Research*, 32(3), 900–904. http://doi.org/10.1016/S0043-1354(97)00234-0

Wu, H., & Patterson, G. K. (1989). Laser-Doppler Measurements of Turbulent-Flow Parameters in a Stirred Mixer. *Chemical Engineering Science*, 44(10), 2207–2221. http://doi.org/10.1016/0009-2509(89)85155-3

Wu, W. (2008). *Computational River Dynamics*. London: CRC Press.

Xiao, H. (2006). Fine Clay Flocculation. In P. Somasundaran (Ed.), *Encyclopedia of Surface and Colloid Science* (5th ed., pp. 2572–2583). Boca Raton, FL: CRC Press.

Yeoh, G. H., Cheung, C. P., & Tu, J. (2014). *Multiphase Flow Analysis Using Population Balance Modeling: Bubbles, Drops and Particles*. Waltham, MA: Butterworth-Heinemann.

Yeung, A. K. C., & Pelton, R. (1996). Micromechanics: A New Approach to Studying the Strength and Breakup of Flocs. *Journal of Colloid and Interface Science*, 184(2), 579–585. http://doi.org/10.1006/jcis.1996.0654

Yuan, Y., & Farnood, R. R. (2010). Strength and Breakage of Activated Sludge Flocs. *Powder Technology*, 199(2), 111–119. http://doi.org/10.1016/j.powtec.2009.11.021

Yukselen, M. A., & Gregory, J. (2004). The Effect of Rapid Mixing on the Break-up and Re-formation of Flocs. *Journal of Chemical Technology and Biotechnology*, 79(7), 782–788. http://doi.org/10.1002/jctb.1056

Yusa, M. (1977). Mechanisms of Pelleting Flocculation. *International Journal of Mineral Processing*, 4(4), 293–305. http://doi.org/10.1016/0301-7516(77)90010-2

Yusa, M. (1987). Pelleting Flocculation in Sludge Conditioning - An Overview. In Y. A. Attia (Ed.), *Flocculation in Biotechnology and Separation Systems* (pp. 755–763). Amsterdam: Elsevier.

Yusa, M., & Gaudin, A. M. (1964). Formation of Pellet-Like Flocs of Kaolinite by Polymer Chains. *American Ceramic Society Bulletin*, 43(5), 402–406.

Yusa, M., & Igarashi, C. (1984). Compaction of Flocculated Material. *Water Research*, 18(7), 811–816. http://doi.org/10.1016/0043-1354(84)90264-1

Yusa, M., Suzuki, H., & Tanaka, S. (1975). Separating Liquids from Solids by Pellet Flocculation. *Journal American Water Works Association*, 67(7), 397–402.

Yuxin, Z., Liang, W., Helong, Y., Baojun, J., & Jinming, J. (2014). Comparison of Sludge Treatment by O3 and O3/H2O2. *Water Science & Technology*, 70(1), 114–119. http://doi.org/10.2166/wst.2014.185

Zhu, Z. (2014). Theory on Orthokinetic Flocculation of Cohesive Sediment: A Review. *Journal of Geoscience and Environment Protection*, 2(5), 13–23. http://doi.org/10.4236/gep.2014.25003

Zlokarnik, M. (2008). *Stirring: Theory and Practice*. Weinheim: Wiley-VCH.

List of publications

Oyegbile, B.; Ay, P.; Narra S., Optimization of Physicochemical Process for Pretreatment of Fine Suspension by Flocculation Prior to Dewatering, *Desalination and Water Treatment*, 2015. doi: 10.1080/19443994.2015.1043591

Oyegbile, B.; Ay, P.; Narra S., Optimisation of Micro-Processes for Shear-Assisted Solid-Liquid Separation in a Rotatory Batch Flow Vortex Reactor, *Journal of Water Reuse and Desalination*, 2015. doi: 10.2166/wrd.2015.057

Oyegbile, B.; Ay, P.; Narra S., Flocculation Kinetics and Hydrodynamic Interactions in Natural and Engineered Flow Systems—A Review, *Environmental Engineering Research*, 2015. doi: 10.4491/eer.2015.086

Oyegbile, B.; Ay, P.; Narra S.; Glaser, C., Apparatus and Method for the Agglomeration of Sludge/Vorrichtung und Verfahren zur Agglomeration von Schlämmen, *German Patent and Trademark Office*, 2016. DE 20 2015 107 682 A1

9 780367 574727